MAN-MADE CARBON DIOXIDE AND CLIMATIC CHANGE:

A Review of Scientific Problems

P.S. Liss

and

A.J. Crane

GEO BOOKS

NORWICH

P.S. LISS

Central Electricity Generating
Board, London, and University of
East Anglia, Norwich

A.J. CRANE

Central Electricity Research
Laboratories, Leatherhead

ISBN hardback 0 86094 140 X
ISBN paperback 0 86094 145 0

published by Geo Books
Regency House
34 Duke Street
Norwich NR3 3AP
England

Printed in Great Britain at the University Press, Cambridge

Contents

	Page
Acknowledgements	v
Preface	iv
PART I The Global Behaviour of Man-Made Carbon Dioxide	1
1 Atmospheric Measurements and Man-Made Inputs	3
Atmospheric measurements	3
Man-made inputs	9
2 Uptake of Carbon Dioxide by the Oceans	13
The equilibrium capacity of seawater to absorb CO_2	13
Exhange of CO_2 across the air-sea interface	18
Oceanic mixing	20
Heterogeneous processes which might increase oceanic uptake of CO_2:	27
a) Biological uptake	27
b) Reaction with marine minerals	28
Experimental approaches to estimating oceanic uptake	29
3 The Terrestrial Biosphere and Atmospheric Carbon Dioxide	31
The possible effect of elevated atmospheric CO_2 on terrestrial photosynthesis	31
The possible effect of change in land use on atmospheric CO_2	33
Possible indicators of the net effect of the terrestrial biosphere on atmospheric CO_2:	36
a) Seasonal variation in atmospheric CO_2	37
b) Isotopes of carbon	39
Conclusions on the role of the biosphere in the anthropogenic CO_2 cycle	41
The airborne fraction	41
4 The Global Budget of Man-Made Carbon Dioxide	43
PART II Climatic Consequences of Increased Carbon Dioxide Concentrations	47
5 The Greenhouse Effect	49
Atmospheric infrared absorption	50
Infrared radiative transfer	52
The direct radiative influence of increased CO_2	54

6 One-dimensional Climate Models 59

Horizontally-averaged vertical column models 59
Latitudinally varying energy balance models 62

7 Global Climate Models 65

Introductions to GCMs 65
The representation of clouds in GCMs 67
The representation of the oceans in GCMs 70

8 Application of Global Climate Models to the Carbon Dioxide Problem 73

CO_2-induced temperature changes 74
CO_2-induced changes in the water budget 78
The transient response to increasing CO_2 concentrations 83
Summary of GCM CO_2 studies 85

9 Clues from the Past 87

Warm world scenarios based on historical analogues 87
Past atmospheric CO_2 concentration deduced from ice cores 88

PART III Implications of Increased Carbon Dioxide for the Use of Energy in the Future 91

Introduction 91
The direct effect of climatic change on energy demand: 92
a) Electricity 92
b) Oil and gas 92
The time-scale of atmospheric CO_2 increase 93
When will it be certain that a CO_2-induced climatic change will occur? 96
Technological options for reducing the rate of increase in atmospheric CO_2 101
Global CO_2 research effort 105

PART IV

Summary 109

References 113

Index 125

Acknowledgements

The authors acknowledge the following who have given permission for the use of their published diagrams as a basis for figures in the text: Drs L. Machta, W.W. Kellogg, R.M. Rotty, H.R. Delcourt, C.A. Hall, W.T. Roach, K. Bryan, J. Hansen, A. Neftel, A. Bjorkstrom and B. Moore; Cambridge University Press, Elsevier Scientific Publishing Company, The Plenum Press, The American Association for the Advancement of Science, MacMillan Journals Ltd, J. Wiley and Sons, D. Reidel Publishing Company and American Journal of Science.

Permission granted via printed policy statements by The American Meteorological Society, The American Geophysical Union and The Optical Society of America to reproduce figures from their journals for use in educational works is also acknowledged, as are the authors whose diagrams in these journals have been used, namely Drs W.S. Broecker, R.A. Hanel, V. Ramanathan, S. Manabe, R.T. Wetherald and A. Lacis.

Preface

Carbon dioxide is the fourth most abundant gaseous constituent of air (after nitrogen, oxygen and argon) and contributes 0.03% to the total volume. Its total amount in the atmosphere is currently about 711 GtC*. The gas plays an important role in controlling the radiation balance and hence the temperature distribution in the atmosphere, through its ability to absorb energy emitted from the earth's surface in the infra-red region. Carbon dioxide is also vital to life processes, being extracted from the atmosphere during photosynthesis and ultimately returned as a result of decomposition and respiration. The amount of CO_2 stored in the atmospheric reservoir is small relative to the amounts of inorganic carbon contained in the oceans (~39 000 GtC, mainly as bicarbonate ions) and in sediments (~20 000 000 GtC, largely as calcium carbonate minerals). Although harder to estimate, the biospheric pool of living and dead organic carbon contains of the order of 2000 GtC, not including the amount in fossil fuels which is thought to be in excess of 7500 GtC.

Prior to the Industrial and Agricultural Revolutions, man's impact on the cycling of carbon between these reservoirs was negligible. With the advent of large-scale industrialisation and agricultural development, human activities have begun to be on a scale potentially large enough to play a significant role in some of these natural cycles. For example, present fossil fuel combustion injects CO_2 equivalent to about 5 GtC into the atmosphere each year, whereas the natural (and balanced) rates of CO_2 exchange between the atmosphere and the biospheric and oceanic reservoirs are both about 100 GtC per annum. Thus, each year fossil fuel burning injects an amount of CO_2 into the atmosphere equivalent to approximately 5% of the atmosphere's two most important natural exchange mechanisms.

---

* GtC = gigatonnes, expressed as carbon

1Gt = 10^9 tonnes = 10^{15} grams

If, as seems likely, human activities which release CO_2 to the atmosphere continue to expand, the atmospheric concentration could increase sufficiently to cause significant changes in global climate within the next century. Estimating the timing and character of a CO_2-induced climatic change depends upon knowledge of the ways in which cycling of carbon within the natural reservoirs determines the atmospheric retention of anthropogenic CO_2, and of the ways in which the climate system responds to increased atmospheric CO_2 levels. These two scientific and technical aspects of what has broadly become known as 'The CO_2 Problem' are treated in detail in this book. Other equally important, though intrinsically less scientific, questions relating to the economic, political and sociological consequences of the possible climatic change are not considered here. (An assessment of these aspects of the problem has been given by Kellogg and Schware (1981)).

The first part of the book is devoted to the global behaviour of anthropogenic CO_2. It deals with the increasing atmospheric concentration, the uptake of CO_2 by the oceans and the complex role of the biosphere as a source and a sink of atmospheric CO_2. The second part deals with the climatic consequences of increased atmospheric CO_2, where most attention is paid to numerical models of climate and their use in studying the sensitivity of climate to increased CO_2. A concluding section addresses some questions related to the CO_2 problem of particular concern to the energy industries.

This book was first produced as an internal report within the Technology Planning and Research Division of the Central Electricity Generating Board (CEGB) and was written jointly at the CEGB Headquarters in London (PSL) and at the Central Electricity Research Laboratories at Leatherhead, Surrey (AJC). Permission to publish the work has been given by the CEGB.

PART I

THE GLOBAL BEHAVIOUR OF MAN-MADE CARBON DIOXIDE

1 Atmospheric Measurements and Man-Made Inputs

Atmospheric measurements

The earliest attempts to measure the concentration of CO_2 in the atmosphere were made well over 200 years ago, although the first results on which any credence should be placed are probably those of Thorpe (1867). These and subsequent measurements made up until the 1950s are suspect for a variety of reasons including unrepresentative nature of samples, contamination during collection, storage and handling, and poor chemical analytical methodology. The early measurements have been critically evaluated by Callendar (1958) and Bray (1959), and the more reliable ones are shown in Fig. 1 which is adapted from Machta (1979). Although there is a lot of scatter in the early data, it is clear why a value of 290 ppm is often assumed to be representative of the CO_2 concentration in the atmosphere prior to large-scale industrialisation. However, this value is by no means generally accepted; for example substantially lower figures (260-270 ppm) have been proposed by Bjorkstrom (1979), Stuiver (1978) and Wigley (1983). This uncertainty is unfortunate, especially so in the context of modelling the increase in atmospheric CO_2 over the last century. Such models can potentially give important information on the sizes of the oceans and biosphere as sinks or sources for anthropogenically produced CO_2, but their utility is reduced if the initial level of CO_2 in the atmosphere is insufficiently constrained.

In spite of the scatter, there is some evidence from the data in Fig. 1 for an increase in atmospheric CO_2 in the first half of this century. Reliable evidence for this trend is provided by accurate measurements made over the last twenty years at several locations. These data are shown in Fig. 2 which gives yearly mean values at Mauna Loa (Hawaii), the South Pole, Point Barrow (Alaska), and for samples collected from Scandinavian aircraft on various commercial routes in the Northern Hemisphere. A striking feature of the results in Fig. 2 is the general agreement between the yearly average measurements made at widely spaced locations, thus establishing that the increase in atmospheric CO_2 is a truly global phenomenon. This geographical

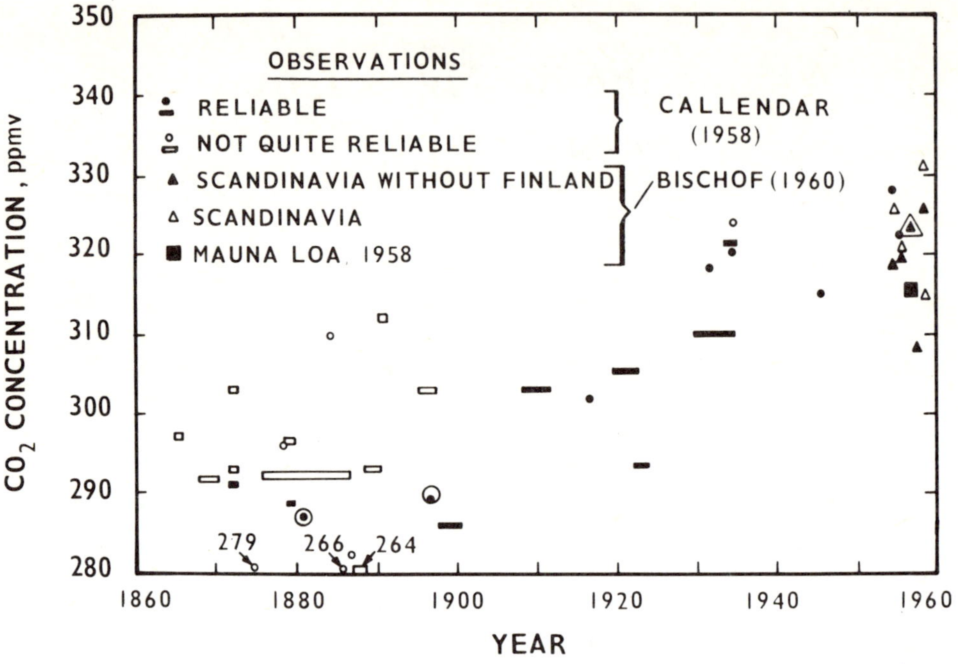

Figure 1. Measured atmospheric CO_2 concentration in the period 1860-1960. (After Machta, 1979)

uniformity in the yearly average values results from the fact that the residence time of CO_2 in the atmosphere is about 7 years so that, irrespective of how and where a CO_2 molecule enters or leaves the system, it is reasonably well mixed within the atmospheric reservoir by the wind systems (annual mean interhemispheric p_{CO_2} difference 1-2 ppm).

Inspection of Figs. 1 and 2 shows an apparent discrepancy between the Scandinavian measurements made in the late 1950s (mean value ~325 ppm) and the beginning of the Mauna Loa record, which starts with a concentration of 315 ppm in 1958. Although this ~10 ppm mis-match has been noted previously (Barrett, 1975), no explanation for it seems to have been given in the literature. Pearman (personal communication) thinks that up to 5 ppm of the discrepancy may arise from systematic differences between the chemical method used to analyse most of the Scandinavian samples and the infrared absorption technique employed for the Mauna Loa samples. The rest must come about from the proximity of Scandinavia to sources of fossil fuel CO_2 release (which might account for another ~1-2 ppm) and sampling problems due to the poor location of some of the collection sites.

The temporal increase shown in Fig. 2 is still continuing; the most recent yearly average value for Mauna Loa being 338 ppm in 1981.

Although the yearly average values always show an increase, the actual magnitude is rather variable. Table 1 gives the yearly increases, for the Mauna Loa data set, as

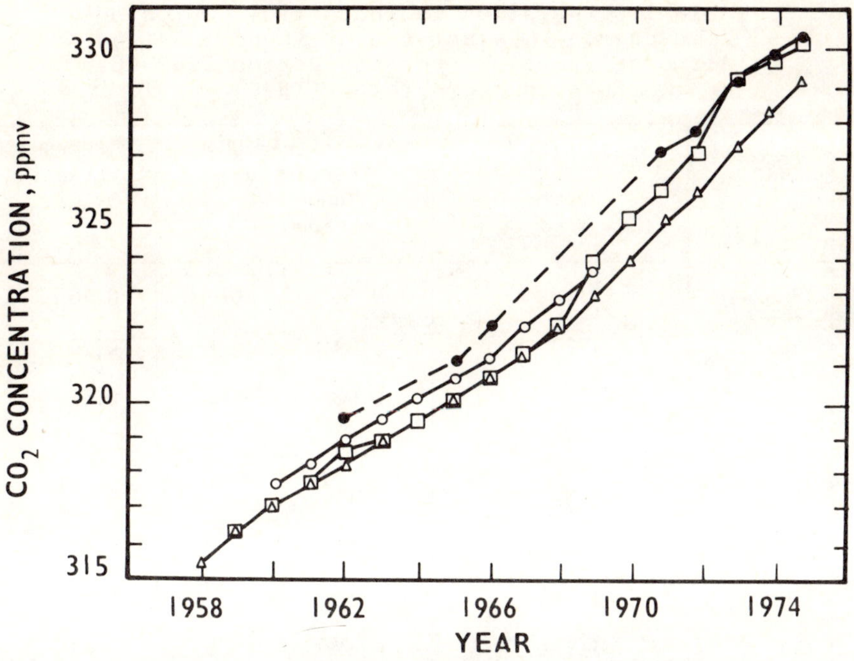

Figure 2. Yearly average atmospheric CO_2 concentrations over the period 1958-1976 at the following places: □ Mauna Loa; △ S. Pole; ● Point Barrow, Alaska; ○ Swedish aircraft flights (N. hemisphere at various heights). (After Kellogg, 1978)

well as values for the 'airborne fraction' (defined as the ratio of the observed yearly atmospheric increase to the total emission that year of CO_2 from the burning of fossil fuel and cement production, assuming these to be the only sources). The airborne fraction thus tells us how much of the CO_2 emitted from the above sources to the atmosphere appears to remain there; the rest having been removed to other environmental reservoirs, of which the oceans appear to be the major one (see Chapter 2).

Variability in the yearly increase in atmospheric CO_2 concentration observed at Mauna Loa, coupled with the steady increase in output, leads to rather large variations in the airborne fraction from year to year, and occasionally to values greater than 1.0. Such variability in the detailed yearly record is not particularly surprising since the emission figures are for the whole world, whereas the measurements in Table 1 are for one location only. If the airborne fraction is calculated for a number of widely spaced stations and a yearly average calculated, the results are found to be more uniform than those for Mauna Loa alone (Machta, 1979). From Table 1 the mean annual airborne fraction for Mauna Loa over the period 1958-1976 is 0.56 ± 0.22. Since the yearly increase in CO_2 concentration differs between observing stations, the calculated airborne fraction will

Table 1: Variation in Industrial CO_2 Output, Atmospheric CO_2 Increase at Mauna Loa, and Airborne Fraction over the Period 1958-76, Adapted from Freyer (1979)

Year	Industrial CO_2 Output ppm*	Atmospheric CO_2 Increase at Mauna Loa ppm	Airborne CO_2 Fraction ‡
1958	1.14	0.61	0.46
1959	1.20	0.89	0.74
1960	1.27	0.68	0.54
1961	1.25	0.86	0.69
1962	1.32	0.46	0.35
1963	1.40	0.60	0.43
1964	1.48	0.62	0.42
1965	1.55	0.67	0.43
1966	1.63	0.78	0.48
1967	1.65	0.88	0.53
1968	1.76	1.89	1.07
1969	1.86	1.32	0.71
1970	2.00	1.04	0.52
1971	2.06	1.15	0.56
1972	2.14	2.19	1.02
1973	2.27	0.54	0.24
1974	2.29	0.51	0.22
1975	2.28	1.08	0.47
1976	2.37	1.58	0.67
1958-76		0.97 ± 0.48	0.56 ± 0.22

* Expressed in ppm CO_2 assuming all the emitted gas remains airborne

‡ Ratio of 3rd to 2nd columns.

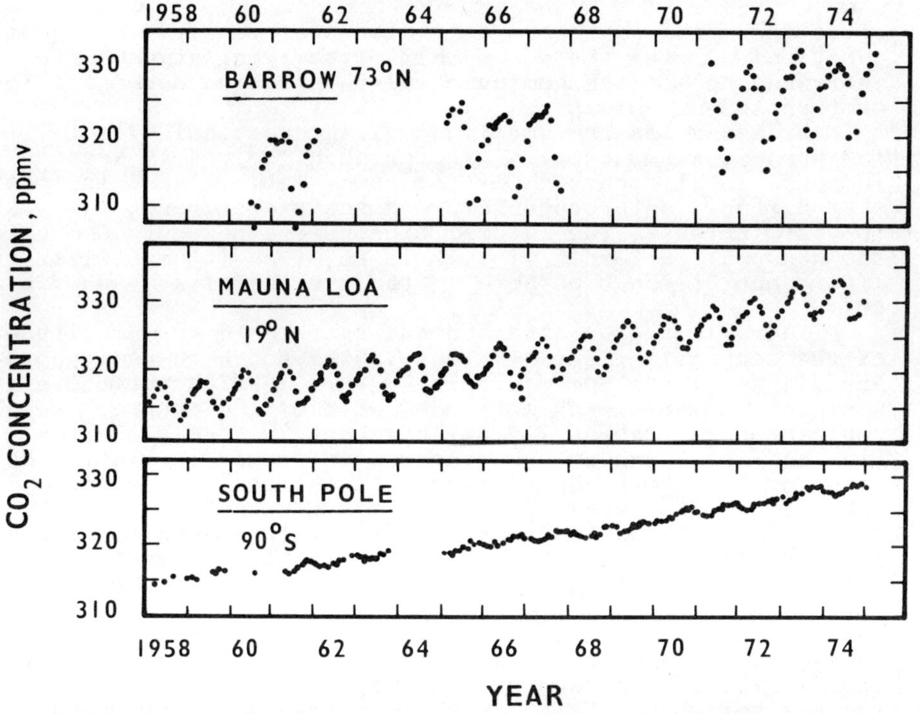

Figure 3. Monthly average atmospheric CO_2 concentrations over the period 1958-1975 at three locations. (After Machta, 1979)

vary geographically. For example, for the period 1959-1978 the mean airborne fraction for Mauna Loa is 0.55 whereas for the South Pole observing station it is 0.51 (Bacastow and Keeling, 1981).

All the above discussion on airborne fractions has been based on the assumption that the sole cause of atmospheric CO_2 increase is fossil fuel consumption with a small (1-2%) contribution from the production of cement. If other net sources of CO_2 to the atmosphere are, or have been, of consequence then the 'true' airborne fraction will not be as calculated (see page 41).

The treatment so far has been in terms of annual average values of atmospheric CO_2. Examination of monthly, rather than yearly, averages highlights some significant seasonal trends. In Fig. 3 such data are given for Mauna Loa together with the less complete records for the South Pole and Barrow, Alaska. There is a clear seasonal pattern in both the Mauna Loa and the Barrow data sets. A similar trend is shown by the South Pole record but it is of much reduced amplitude. Maxima and minima on the yearly curves seem to correspond approximately to the local end of winter respiration/decomposition and summer plant growth, respectively. For example, at Mauna Loa the decline in CO_2 from about May to October is thought to be due to net consumption of the gas by terrestrial plants in that latitude band, since then photosynthesis dominates over respiration/decomposition. During the other half of the year atmospheric CO_2 increases due to the dominance of respiration/decomposition over photosynthesis.

At Mauna Loa the amplitude of the seasonal effect (>3 ppm) is substantially greater than the yearly average increase in atmospheric CO_2 (~1 ppm - Table 1). This implies that although anthropogenic processes probably set the year to year increase, the natural biosphere appears to dominate the seasonal pattern. If this is so, then it emphasizes the importance of the biosphere in the cycling of atmospheric CO_2.

There is a clear latitudinal pattern in the amplitude of the seasonal effect. Values in the southern hemisphere are all small (<1 ppm e.g. Fig. 3 for S. Pole) but there is a steady increase with latitude north of the equator, reaching amplitudes as high as 7-8 ppm at about 70°N (e.g. Barrow, Fig. 3). Although no observations exist, it is likely that even further north the amplitude of the seasonal effect will tend to decrease again.

Machta (1972) has been able to explain much of this latitudinal variation by use of a model incorporating only atmospheric diffusion and estimates of the uptake/release of CO_2 by the terrestrial biosphere in broad (20°) latitude bands. Thus the small amplitude of the seasonal effect in the southern hemisphere is due to the small extent of biospherically active land surfaces south of the equator. In the northern hemisphere, with its substantial vegetated land area, there are two conflicting trends. Although at low latitudes biological activity is high it is not much dependent on time of year. In contrast, the seasonal effect is very marked at high latitudes but the growing season is so short that little carbon cycles through the system. These two effects combine to give a maximum seasonal variation in atmospheric CO_2 in the latitude band 50-70°N.

The fact that the above model, containing the land biosphere as the only source/sink, explains much of the variation within the annual cycle in atmospheric CO_2 implies that other environmental reservoirs (e.g. the oceans) play only a small part in the seasonal cycling of CO_2. Machta (1972) argues that the lack of importance of the oceans in this context comes about because the effects of seasonal change in ocean water temperature and biological activity are in opposition and so tend to cancel each other out. For example, in the spring and summer increases in the temperature of the surface water will reduce its ability to hold CO_2 (see Fig. 8) and so the tendency will be for gas to be lost to the atmosphere. However, photosynthetic activity in the surface water will work in the opposite sense by extracting CO_2 and converting it to carbohydrate etc.

From the above discussion of atmospheric measurements it is clear that in order to understand past trends in atmospheric CO_2 and to predict changes in the future the following three factors are of paramount importance: (i) Anthropogenic inputs from fossil fuel consumption and cement manufacture, (ii) Uptake of excess CO_2 by the oceans, (iii) Possible changes in the terrestrial biosphere resulting from man's intervention. The first of these is dealt with below and the other two are the subjects of Chapters 2 and 3 respectively.

Man-made inputs

A major source of man-made CO_2 for the atmosphere is the combustion of fossil fuel. The contribution from this source for the period since 1860 was estimated by Keeling (1973a) using United Nations data on the amounts of different types of fuel burned and a knowledge of the quantity of CO_2 produced from the combustion of unit amounts of each fuel type. Rotty (1973, 1974) revised the calculations to include an additional source since 1950 from the flaring of natural gas and the small contribution from the manufacture of cement. Subsequently the calculations have been updated regularly by Rotty (e.g. Rotty, 1977, 1979, 1982).

The results of the exercise are shown in Fig. 4 which clearly shows the expected sharply rising trend of CO_2 output with time. However, when the emissions are plotted on a logarithmic scale (Fig. 5) it becomes apparent that the rate of increase in CO_2 injected into the atmosphere has, with well defined and explicable exceptions, been remarkably constant. The exceptions correspond to the periods of the two World Wars and the 'Great Depression' of the 1930s. These apart, the rate of increase has been close to 4.3% per year from 1860 to the early 1970s. If the periods of lesser fuel usage are included then the average annual increase in the 118 years from 1860 to 1978 has been 3.5%, with total emissions during this time being 155 GtC.

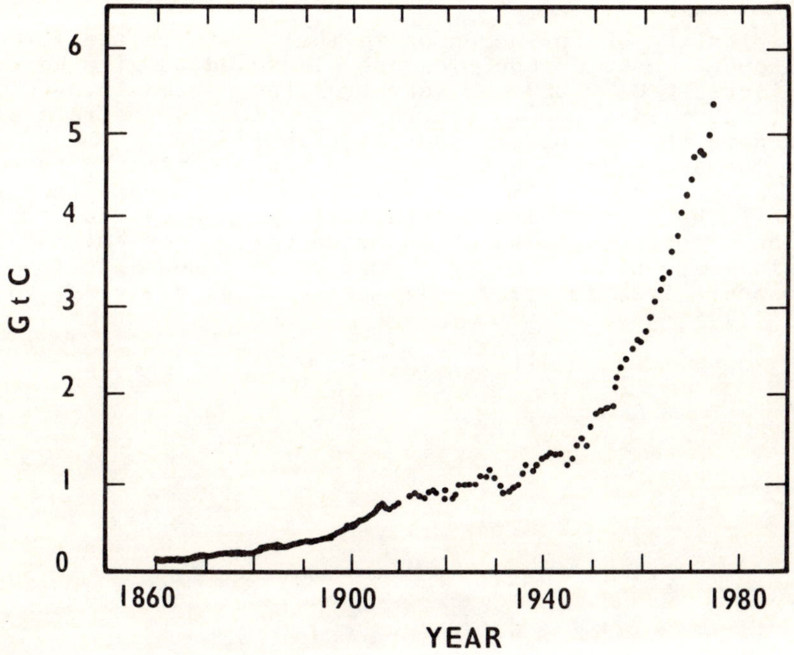

Figure 4. Annual production of CO_2 from fossil fuels and cement 1860-1977 (1860-1949 data from Keeling, 1973a). (From Rotty, 1980)

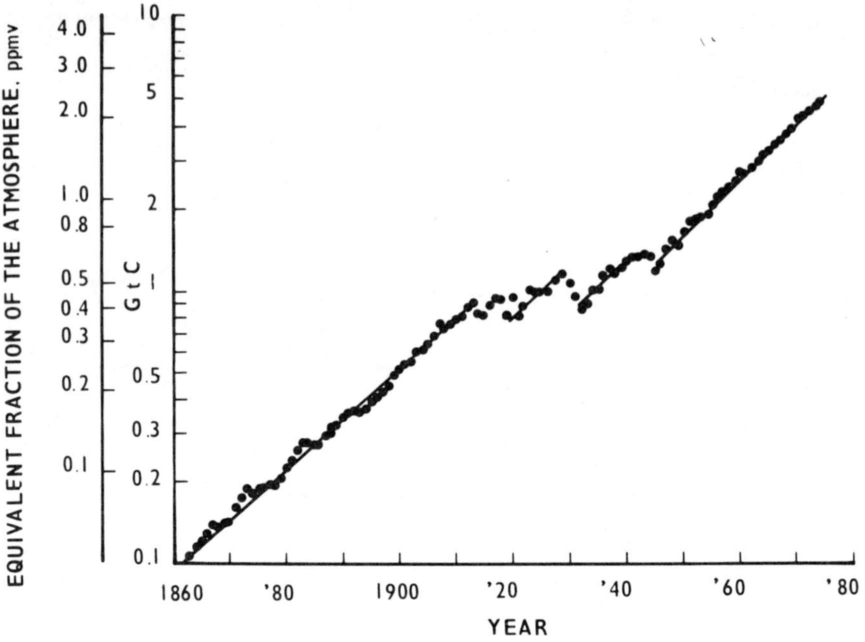

Figure 5. Annual production of CO_2 from fossil fuels and cement 1860-1974 (1860-1949 data from Keeling, 1973a). (From Rotty, 1977)

With the sharp increases in fuel costs in the 1970s, anthropogenic CO_2 production has slackened to give an average increase of 2.25% per year over the period 1973-80 (Fig. 6), mainly due to the rapid decline in the rate of growth in the use of oil and gas (Fig. 7).

The above data on CO_2 emissions are estimated to be accurate to about ±13%. As will become apparent, they are clearly the most reliable quantitative information available in the global budgeting of man-made CO_2. Some estimates of possible amounts of CO_2 which may be injected into the atmosphere from future fossil fuel burning are given in Part III.

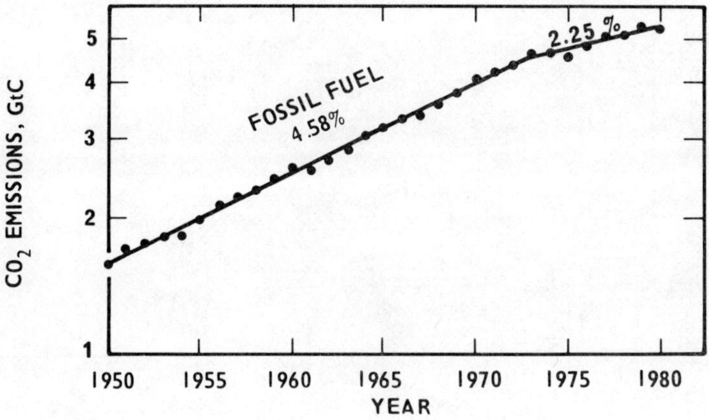

Figure 6. Annual global production of CO_2 from fossil fuels and cement, with indicated growth rate. (From Rotty, 1983)

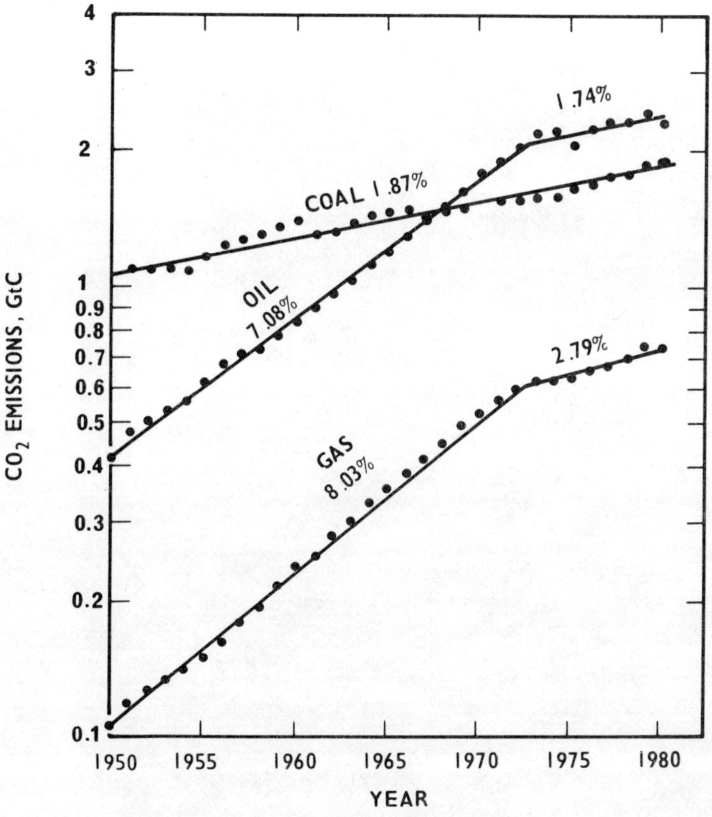

Figure 7. Annual CO_2 emissions from each type of fossil fuel, with growth rates. (From Rotty, 1983)

2 Uptake of Carbon Dioxide by the Oceans

At present, estimation of the magnitude of the oceanic sink for man-made CO_2 is generally done by use of models (other possible approaches involving more direct environmental measurements are outlined on page 29. The models involve basically three stages:

(i) Estimation of the equilibrium capacity of seawater to take up excess atmospheric CO_2, (ii) Quantification of the rate at which such excess CO_2 can cross the air-sea interface into the surface oceans, (iii) Assessment of the rate at which surface waters mix with deeper layers. These various stages are discussed in turn below.

The equilibrium capacity of seawater to absorb CO_2

The following equations describe the uptake of CO_2 by pure water:

$$CO_{2\,(g)} \underset{}{\overset{H_2O}{\rightleftharpoons}} CO_{2\,(aq)} \qquad \ldots(1)$$

$$CO_{2\,(aq)} + H_2O_{(l)} \rightleftharpoons H_2CO_{3\,(aq)} \qquad \ldots(2)$$

$$H_2CO_{3\,(aq)} \rightleftharpoons H^+_{(aq)} + HCO^-_{3\,(aq)} \qquad \ldots(3)$$

$$HCO^-_{3\,(aq)} \rightleftharpoons H^+(aq) + CO^{2-}_{3\,(aq)} \qquad \ldots(4)$$

Equation (1) and its associated equilibrium constant quantify the ratio of CO_2 molecules in the liquid to those in the gas phase. Similar expressions exist for the dissolution of all gases and the numerical value of the equilibrium constant for each reaction is a measure of the solubility of the gas in water (often called the Henry's Law constant). For many gases (e.g. O_2, N_2, CO, CH_4, N_2O) this equilibrium constant is all that is required to describe their gas/liquid partitioning since the gases do not react chemically with the water. For CO_2 and other gases which are reactive in the aqueous phase additional equations are necessary. In the case of dissolution of CO_2, $H_2CO_{3\,(aq)}$ is formed (equation (2)) which although a weak acid can dissociate to produce $HCO^-_{3\,(aq)}$ and $CO^{2-}_{3\,(aq)}$ ions with release $H^+_{(aq)}$ ions (equations (3) and (4)).

Seawater is, of course, not pure water since it contains substantial amounts of dissolved salts (about 35 g l^{-1} equivalent to an ionic strength of ~0.7 mol). Much of the dissolved material is neutral electrolyte (Na^+, K^+, Cl^-, SO_4^{2-}) and although this alters somewhat the values of the various equilibrium constants in the system described by equations (1) to (4) the effects are not very large. Of more importance in the present context is the fact that seawater also contains HCO_3^- and CO_3^{2-} ions, derived not from dissolution of atmospheric CO_2 but arising as a result of the weathering of non-neutral salts (such as $CaCO_{3(s)}$), and these constitute its natural alkalinity. The weathering of such minerals occurs either on land with subsequent transport of the resulting ions via rivers into the oceans, or from the dissolution of suspended particles and sea-bed sediments in the marine environment. The natural alkalininty can be measured experimentally by titration of the seawater against strong acid (HCl is generally used) and is then called Titration Alkalinity (T.A.), which is defined as follows:

$$\text{T.A.} = [HCO_3^-] + 2[CO_3^{2-}] + [H_2BO_3^-] + ([OH^-] - [H^+]) \quad \ldots(5)$$

where the square brackets indicate equilibrium concentrations in mol l^{-1} (or mol kg^{-1}), and T.A. is in equiv. l^{-1}. The role of $H_2BO_3^-$ and other weak acid anions in contributing to the T.A. of seawater is small (<5%) and for simplicity will be ignored in all the subsequent discussion. With this simplification titration alkalinity is more correctly termed carbonate alkalinity (C.A.).

Another parameter often used in describing the carbonate chemistry of natural waters is Total Inorganic Carbon content (ΣC) which is defined as follows:

$$\Sigma C = [CO_{2(aq)}] + [H_2CO_3] + [HCO_3^-] + [CO_3^{2-}] \quad \ldots(6)$$

where square brackets indicate concentrations in mol l^{-1} (or mol kg^{-1}).

One effect of the alkalinity-producing ions is to raise substantially the pH of the seawater above the value of a solution of similar ionic strength but containing only neutral electrolyte and in equilibrium with the same atmospheric partial pressure of CO_2. For example, a solution of neutral electrolyte of ionic strength 0.7 mol in equilibrium with a gas phase CO_2 partial pressure of 0.0003 atm (i.e. close to the present level in air) would have a pH of about 5.4, as against 8.2 for a water of similar ionic strength and in equilibrium with the same gas phase but having a carbonate alkalinity of 2.4×10^{-3} equiv l^{-1} (a typical value for seawater).

An important effect of the alkalinity of seawater, more especially the presence of CO_3^{2-} ions, which constitute about 15% of the total carbonate alkalinity, is to increase the seawater's ability to absorb excess atmospheric CO_2. Essentially carbonate ions contributing to the alkalinity will react with H^+ ions derived from dissociation of H_2CO_3 (equations (3) and (4)) to form HCO_3^- ions. This effectively removes free H^+ ions and the system will respond by dissolving more CO_2 from the atmosphere in order to make good the short fall. To gain an idea of the size of this effect, we have calculated that for present levels of atmospheric CO_2 a seawater with C.A. = 2.4 x 10^{-3} equiv l^{-1} has about eight times the capacity to absorb excess CO_2 as the same water but with its alkalinity-producing ions replaced by neutral electrolyte.

In ocean uptake models the parameter most commonly used to quantify the static (i.e. equilibrium) ability of seawater to take up excess CO_2 is called the buffer (or Revelle) factor. It is the ratio of the fractional increase in atmospheric CO_2 to the fractional increase this produces in the total inorganic carbon content of seawater. There are several ways in which the buffer factor (B_H) can be formulated mathematically - here we will use the following definition:

$$B_H = \frac{\delta p_{CO_2}/p_{CO_2}}{\delta\Sigma C/\Sigma C} \qquad \ldots(7)$$

where the deltas refer to small changes in the partial pressure of CO_2(p_{CO_2}) and ΣC. A typical value for the present atmosphere/ocean system is 10, which for example means that a 10% increase in p_{CO_2} will produce only a 1% increase in the total inorganic carbon content of seawater once equilibrium has been achieved. B_H as defined here applies strictly only for infinitely small changes since addition (or subtraction) of CO_2, or any of the other inorganic carbon species in the air-seawater system, will lead to a different numerical value for the buffer factor. Fortunately, as will become apparent (see Fig. 8), B_H is not strongly dependent on p_{CO_2} so this is not an important restriction unless very accurate values are required.

In order to see how B_H and other parameters vary with increase in atmospheric CO_2 we have performed equilibrium calculations using the relationships developed by Park (1969) together with CO_2 solubility data from Weiss (1974) and dissociation constants for carbonic acid from Hansson (1973). Again borate and other non-carbonate alkalinity components have been ignored. The parameters CO_2(aq) (strictly CO_2(aq) + H_2CO_3(aq) but the concentration of aquated CO_2 is always much greater than that of undissociated carbonic acid), HCO_3^-(aq), CO_3^{2-}(aq), pH and B_H have been calculated for preset fixed values of p_{CO_2} starting at 300 ppm (close to the pre-industrial level) and increasing in steps of 300 ppm up to 3000 ppm. Although this is an unrealistically large range for p_{CO_2}, it does help to show the changes which take place in the seawater carbonate system as atmospheric carbon dioxide increases. Here B_H, which is calculated as the mean value between p_{CO_2} steps, will be somewhat inaccurate due to the rather large increments but, for the reason given earlier, this should not hide any important trends. In order to see the effects of temperature the calculations have been performed at both 10° and 25°C. The results are shown in Fig.8.

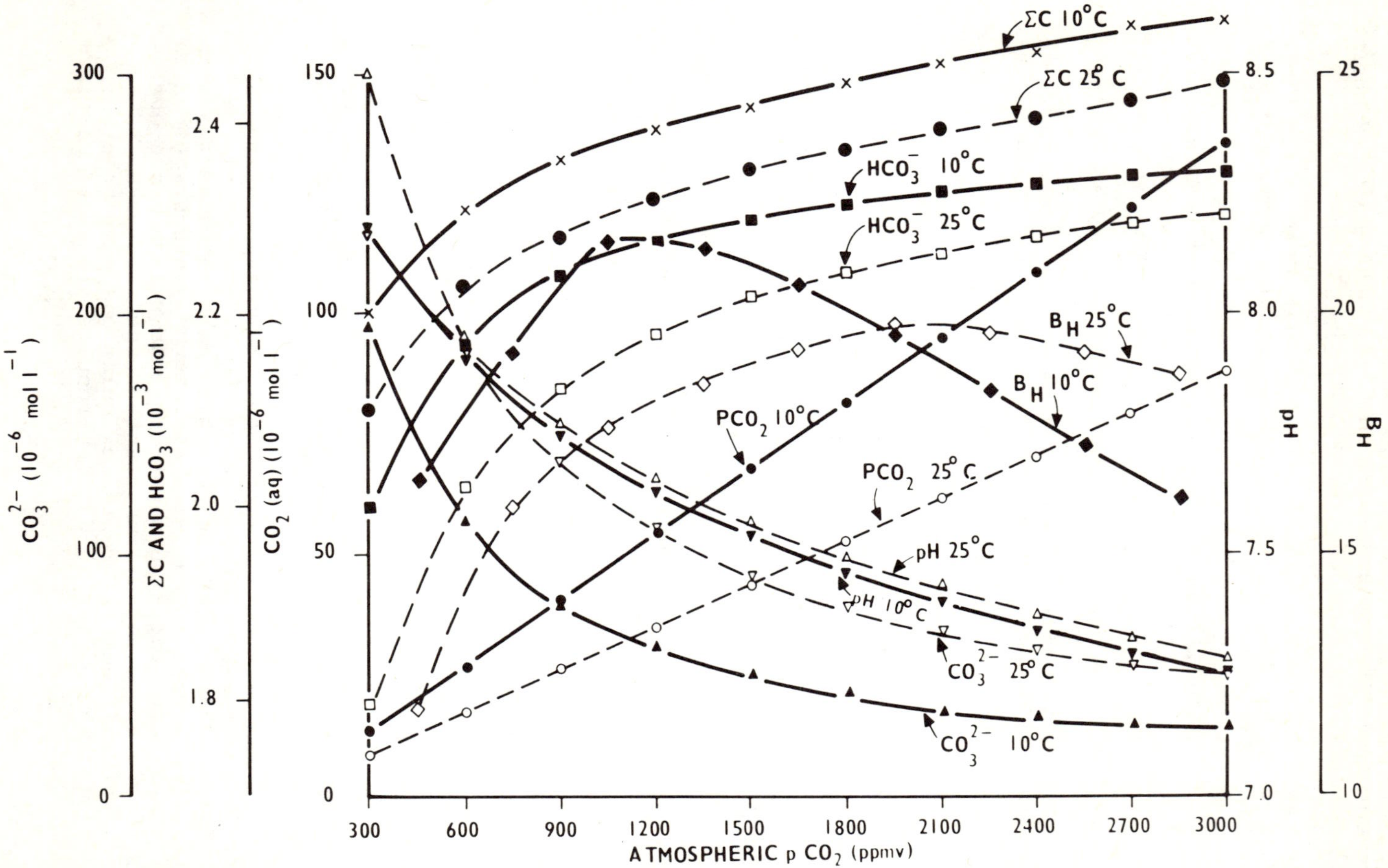

Figure 8. Variation of inorganic carbon speciation, pH and B_H (buffer factor) with p_{CO_2} for a seawater of C.A. 2.4 x 10^{-3} equiv l^{-1} (no borate) and salinity of 35 ‰ for temperatures 10 and 25°C

Important trends apparent from Fig. 8 are as follows:

$CO_2(aq)$ This species shows a linear increase with p_{CO_2} as expected for dissolution of an ideal gas in water, i.e. Henry's Law is obeyed. Since the dissolution reaction is exothermic ($\Delta H^o = -19.4$ kJ mol^{-1}) the amount of dissolved CO_2 is always greater at 10° than at 25^oC.

$HCO_3^-(aq)$ Bicarbonate is the dominant aqueous form of inorganic carbon under all conditions covered by Fig. 8. Its concentration increases with p_{CO_2} but at a rate which decreases as the partial pressure of CO_2 rises. Bicarbonate is always higher at 10^oC than at 25^oC.

$CO_3^{2-}(aq)$ In contrast to $HCO_3^-(aq)$, the concentration of $CO_3^{2-}(aq)$ decreases as p_{CO_2} rises. As discussed earlier this is because addition of CO_2 to the water leads to the production of H^+ ions which convert CO_3^{2-} to HCO_3^-, thereby giving seawater its substantial ability to take up CO_2 from the atmosphere. In the case of $CO_3^{2-}(aq)$ the concentration at 10^oC is always lower than that at 25^oC (contrast $HCO_3^-(aq)$).

pH The pH of seawater drops as p_{CO_2} increases, essentially because the natural alkalinity of the water is being neutralized by the carbonic acid formed from the incoming CO_2. It is interesting to note that doubling p_{CO_2} from its pre-industrial level of about 300 ppm to 600 ppm reduces the pH of the seawater by about 0.3 units, which is approximately the size of natural variations found in the oceans due to biological uptake/release or spatial inhomogeneity. Thus, increases in atmospheric CO_2 above 600 ppm will lead to a mean seawater pH outside the range currently observed. Possible effects of such a pH decrease on marine organisms are discussed by Knutzen (1981).

B_H The buffer factor at both temperatures begins by increasing with atmospheric CO_2 but subsequently goes through a maximum and then decreases at high values of p_{CO_2}. At 10^oC the maximum occurs when p_{CO_2} is between 900 and 1200 ppm when $B_H = 21.5$. For 25^oC the maximum value of B_H is 19.7 which is achieved when p_{CO_2} is between 1800 and 2100 ppm. Similar maxima were predicted by Takahashi (1979) and his values are in reasonable agreement with those given in Fig. 8.

At first sight the maximum and subsequent decrease in B_H of high p_{CO_2} may be surprising. We have argued that much of the ability of seawater to take up CO_2 is due to reaction between the carbonic acid formed and the pool of CO_3^{2-} ions arising from its natural alkalinity. Since it is clear from Fig. 8 that the concentration of CO_3^{2-} ions decreases continuously with increase in p_{CO_2}, one might predict that this would lead to a monotonic increase in B_H. That this is not the case arises from the way that B_H is defined in equation (7). In essence, with increasing p_{CO_2}, the numerator in this expression decreases so tending to reduce B_H, whereas the corresponding decrease in the denominator tends to increase B_H. However, the rate of change of the denominator at low p_{CO_2} exceeds that of the numerator so B_H increases. At high p_{CO_2} the situation is reversed leading to the maximum value of B_H at some intermediate value of p_{CO_2}. Other mathematical formulations of the buffer factor, which relate changes in p_{CO_2} and ΣC to their pre-industrial values, $(p_{CO_2})_o$ and $(\Sigma C)_o$ (e.g. those used by Siegenthaler and Oeschger (1978) and Fisher and Macqueen (1980)) do not exhibit the above maximum but show an increase with rise in atmospheric carbon dioxide which approaches an asymptotic value $(\Sigma C)_o/(p_{CO_2})_o$. It should be stressed that, however the buffer factor is defined, the uptake of CO_2 by seawater must always decrease as p_{CO_2} increases. It is apparent from Fig. 6 that for p_{CO_2} values less than about 1800 ppm the buffer factor at 10^oC is numerically greater than that at 25^oC, but with p_{CO_2} above 2100 ppm the situation is reversed.

Here we have been concerned only with the equilibrium (static) capacity of seawater to take up excess CO_2. Although the predictions of such a treatment will indicate the ultimate state of the atmosphere and ocean with respect to CO_2, the time necessary for the equilibrium to be achieved will be determined by any kinetic constraints imposed on the system. For example, if the whole ocean equilibrated with the atmosphere the man-made CO_2 would be partitioned 85% in the ocean, 15% in the air. In the case of oceanic uptake of CO_2 the two potential kinetic barriers to the attainment of equilibrium are the rate of exchange of the gas across the sea surface and the rate at which CO_2 entering surface waters is mixed down into the main body of the oceans. These kinetic constraints are the subject of the next two sections. We have already seen that, although there are measurable geographical and temporal changes in atmospheric CO_2, there is considerable correspondence between increases at widely spaced locations, so that from the point of view of oceanic uptake the atmosphere may be considered as a homogeneous reservoir.

Exchange of CO_2 across the air-sea interface

Near the ocean surface the water is generally well mixed by the action of wind, waves and tides; the depth of this upper layer being on average about 75m. It is separated from the deep oceans by a region approximately 1km thick where the water temperature decreases rapidly with depth (called the thermocline). The temperature gradient in the thermocline region makes it stably stratified and hence an important barrier to downwards transfer of excess CO_2 taken up by the upper mixed layer (see page 20). Since the surface layer is well mixed, in CO_2 uptake maodels it is generally treated as being homogeneous with the rate at which it responds to a change in the atmospheric partial pressure of CO_2 determined by the rate at which the gas can cross the air-sea interface.

Gas exchange rates across natural air-water interfaces are often specified in terms of a transfer velocity, which is defined as the flux of gas per unit of interfacial concentration difference. Although the global mean value for the transfer velocity is reasonably well known (for example, the Radon Deficiency method as applied to the GEOSECS cruises in the Atlantic and Pacific oceans predicts an average CO_2 transfer velocity of 16.2cm h^{-1}, which agrees quite well with the value of 18.7cm h^{-1} deduced from the distribution of natural $^{14}CO_2$ (Peng et al., 1979)), on more local time and space scales it is rather poorly quantified. It seems intuitively obvious that the transfer velocity ought to be related to a measure of surface water turbulence, such as wind speed, and clear relationships are indeed observed in laboratory wind tunnel studies (see, for example, Liss (1983) for a compilation of results). However, the only detailed field results presently available (from the GEOSECS cruises) show little, if any, relationship between transfer velocity and wind speed (Hasse and Liss, 1980). This is obviously an important discrepancy if one wishes to integrate local partial pressure differences into a global air-sea flux (see page 29), or to quantify directly the transfer in specific sink areas such as the region around 70°N in the Atlantic or in Antarctic waters. Notwithstanding such difficulties, the global value of the transfer velocity seems reasonably well established at between 15 and 20cm h^{-1} - here we will use the lower value.

Gas exchange across an air-water interface is a first order process so that the time response of the aqueous phase to a change in gas phase partial pressure will be exponential. To obtain some feeling for the response time of the oceanic mixed layer to a change in the interfacial concentration difference the above global value for the transfer velocity may be used. We calculate that for a step change in atmospheric concentration of an inert gas the 75m of mixed layer will reach half its new equilibrium concentration in about 15 days. However, this calculation applies only to the transfer of gas in unassociated form in the water, in the present case $CO_2(aq)$. As shown earlier, CO_2 entering the oceans will also react with CO_3^{2-} ions in the water, and this leads to the uptake of about eight times more CO_2 and a concomitant increase in the response time of the mixed layer (i.e. 15 x 8 = 120 days or about 1/3 of a year). Broecker et al. (1980b) have shown that for radioactively labelled $^{14}CO_2$ the response time is longer again by a factor of B_H (i.e. about 10) since, in order for radioactive equilibrium to be achieved, not only CO_3^{2-} but also HCO_3^- has to be re-equilibrated.

The important conclusion from this discussion is that the half-time for the oceanic mixed layer to equilibrate with stable $^{12}CO_2$ in the atmosphere is about a third of a year. Thus passage of CO_2 across the air-sea interface is not a major constraint on uptake of the gas by the oceans and the mixed layer will respond rather rapidly to changes in atmosphere CO_2 concentration. However, despite this rapid equilibration, the mixed layer, because of its small volume, contains only about 18% of the excess CO_2 taken up by the oceans to this time (Broecker et al., 1979).

Oceanic mixing

It has been demonstrated above that owing to its small volume the surface mixed layer of the ocean is not in itself a very important sink for excess carbon dioxide. The total uptake by the oceans depends on the extent to which these surface waters can mix with the intermediate (thermocline) and deep ocean layers beneath, containing about 15 and 50 times as much carbon, respectively, as the surface layer.

Generally the oceans are stably stratified and only when exceptionally dense water is formed at the surface, following cooling or salinity increase due to ice formation, can sinking to appreciable depths occur. The Norwegian and Greenland Seas and around Antarctica are the principal regions where these processes are effective in producing deep ocean water. This sinking is balanced by slow, upward motion of deep water in other regions of the ocean.

Convective mixing between the surface and intermediate water (75 - 1000m) occurs over a wider area. Wintertime cooling of the surface initiates overturning in temperate latitudes, favoured regions being the poleward sides of the major currents in the North Atlantic and North Pacific and in the circumpolar convergence zone in the Southern Ocean.

Outside the regions of convective overturning, vertical profiles of oceanic tracers in the thermocline have often been interpreted as evidence for local vertical diffusive mixing with the surface water. However careful analyses of bomb-produced ^{14}C, ^{3}H and ^{3}He (e.g. Rooth and Ostlund, 1972; Jenkins, 1980) have demonstrated that, at least in the large anticyclonic subtropical gyres, vertical diffusion plays only a minor role in the vertical redistribution of tracers. An alternative, and probably dominant, mechanism by which water can be exchanged between surface and intermediate layers is lateral mixing along isopycnal surfaces, i.e. surfaces of constant potential density. The wind-driven circulation ensures that isopycnal surfaces lying at depths of several hundred metres in the tropics outcrop in the mixed surface layer in temperate latitudes (as shown in Fig. 9). In this way tracers taken up in temperate surface waters are transported to the tropical thermocline. Mesoscale eddies capable of accomplishing lateral mixing have been observed widely in the last decade. Another possibly important mechanism for lateral exchange, involving mass transport, has been proposed by Stommel (1979). He suggests that during winter and spring, downward Ekman pumping in the anticylonic subtropical gyres transfers water from the surface layers to deeper layers at lower latitudes. The strength of this particular process may lead to substantial transfers of heat and carbon out of the surface ocean.

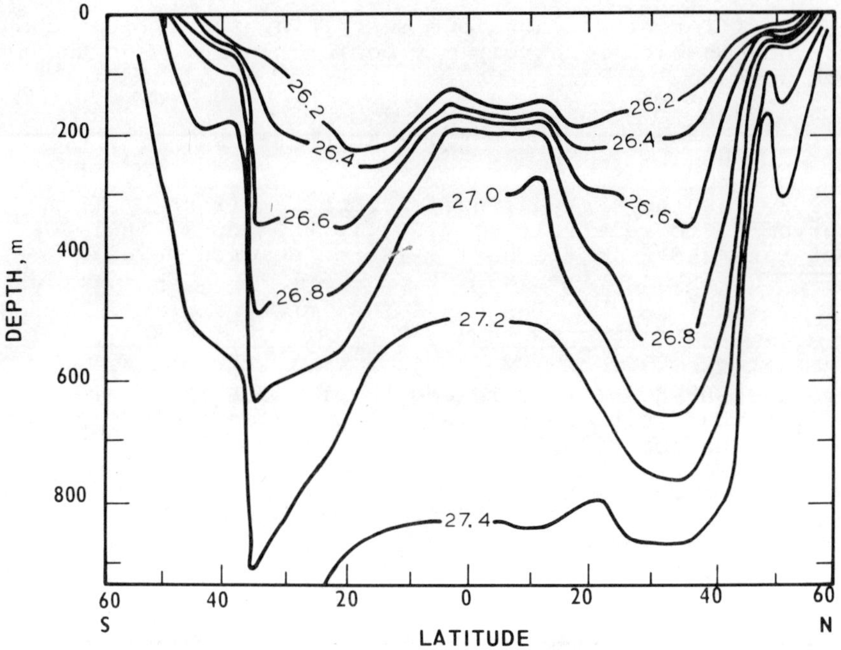

Figure 9. Isopycnal (constant potential density) surfaces as a function of depth along the Western Atlantic GEOSECS track. Values shown are σ_θ (e.g. 26.2 indicates a potential density of 1026.2kg m^{-3}). (After Broecker, et al., 1978)

The accumulation of fossil fuel carbon in the oceans is as yet insufficient for direct observation using the techniques currently available (see page 29). Thus models have been developed to determine the fraction of carbon dioxide emissions to date which has been absorbed in the oceans and to estimate how the oceans will respond to future emissions. Most are analogue models in the sense that they were designed and calibrated to reproduce the observed distributions of natural and bomb-produced tracers, rather than being based on the equations of motion, thermodynamics and continuity, as are the atmospheric circulation models.

The availability of data suitable for calibration and validation has been inadequate for all but the most elementary models. Since these have been used widely we shall discuss the principles on which they are based. After assessing their performance we shall go on to illustrate the ways in which more advanced models are beginning to emerge from this fundamental base.

The simplest models (e.g. Craig, 1957; Keeling 1973b) are those in which the atmosphere and ocean are treated as well- or randomly-mixed reservoirs. The transfer of carbon between two reservoirs is then proportional to the total amount of carbon in the upstream reservoir. Fig. 10 shows a model of this type incorporating a two-box ocean. Denoting the preindustrial carbon contents of the reservoirs by N_a, N_m, N_d and the time-dependent departures therefrom by n_a,

n_m, n_d, differential equations describing the rate of change of carbon in each reservoir may be formulated. For the atmosphere we have

$$\frac{dn_a}{dt} = - k_{am}(N_a + n_a) + k_{ma}(N_m + \xi n_m) + \gamma_f(t) \qquad ...(8)$$

where k_{ij} is the exchange coefficient for transfer from reservoir i to reservoir j, $\gamma_f(t)$ is the fossil fuel CO_2 input rate and ξ is the buffer factor defined here as

$$\frac{(p_{CO_2}) - (p_{CO_2})_o}{(p_{CO_2})_o} \Bigg/ \frac{\Sigma C - (\Sigma C)_o}{(\Sigma C)_o} \qquad ...(9)$$

where the subscript zero refers to preindustrial values. Assuming that in preindustrial times a steady state prevailed, equation (8) yields

$$k_{am}N_a = k_{ma}N_m \qquad ...(10)$$

so that one of the exchange coefficients can be eliminated. Equations similar to (8) and (10) apply to the mixed layer and deep ocean reservoirs.

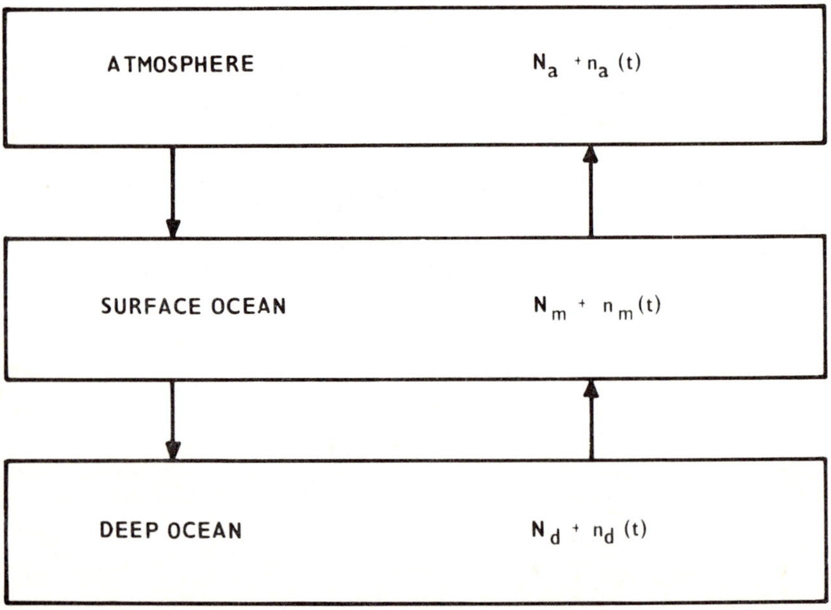

Figure 10. A simple three-reservoir model incorporating a two-box ocean. N_x denotes preindustrial carbon content of reservoir x and $n_x(t)$ the time-dependent departure therefrom.

Calculations, from observations, of the average natural ^{14}C concentrations in the atmosphere ($^{14}N_a$), mixed layer ($^{14}N_m$) and deep ocean ($^{14}N_d$) allow the inter-reservoir exchange coefficients to be evaluated from an assumed natural ^{14}C balance. Again using the atmospheric equation as an example, we have

$$0 = -k_{am}\,{}^{14}N_a + k_{ma}\,{}^{14}N_m - \lambda\,{}^{14}N_a + {}^{14}\Gamma_a \qquad \ldots(11)$$

where λ is the ^{14}C decay constant and $^{14}\Gamma_a$ the natural atmospheric source of ^{14}C. Similar equations, excluding the source term, apply to the mixed layer and deep ocean reservoirs.

The set of three linear, first order equations (equation (8) and corresponding equations for n_m and n_d) may be solved to give the carbon content of each reservoir as a function of time. If $\gamma_f(t)$ is approximated by an exponential increase, the time dependence drops out and the resulting algebraic equations may be solved analytically to yield the time-independent partitioning of fossil fuel carbon amongst the reservoirs.

Since in this chapter we are concerned only with the oceanic uptake of CO_2, the biosphere has been assumed to play a neutral role in the model described. However, it would be straightforward to include an additional reservoir exchanging carbon with the atmosphere if it were wished to model the possible response of the biosphere to increased CO_2 concentrations or the change in biomass resulting from land-use changes, as discussed in Chapter 3.

If the two-box ocean model described is calibrated against observed surface and deep ocean ^{14}C observations, it is possible to reproduce the airborne fraction calculated for the period since 1957 (~0.53) only if the surface ocean mixed layer reservoir is made many times deeper than the mixed layer of the real ocean. For a 75m mixed layer depth, the model gives an airborne fraction in excess of 0.8 (Bacastow and Bjorkstrom, 1981).

Attempting to improve the uptake by the deep ocean while retaining a realistic mixed layer depth, Oeschger et al. (1975) replaced the well-mixed deep ocean box with a diffusive reservoir, allowing a vertical gradient of carbon to develop in the deep ocean. The vertical eddy diffusion coefficient was determined from observed profiles of natural ^{14}C in the deep oceans. This box-diffusion model is better able to accommodate fossil fuel CO_2, but still produces an airborne fraction around 0.65*, i.e. significantly in excess of that observed.

---

* In the original paper by Oescheger et al. (1975) an erroneously small value for the preindustrial mixed layer carbon content gave lower airborne fractions than when correctly calibrated (see Bacastow and Bjorkstrom, 1981).

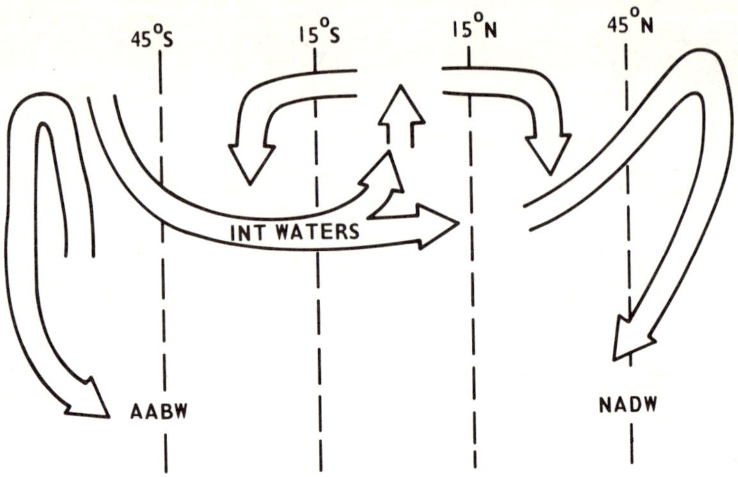

Figure 11. Schematic circulations for the Atlantic Ocean showing the formation of intermediate waters, Antarctic Bottom Water (AABW) and North Atlantic Deep Water (NADW). (After Broecker et al., 1980b)

If the ocean models are correct in predicting a lower uptake capacity than suggested by the observed airborne fraction, the conclusion is that other sinks for fossil fuel CO_2 exist. On the other hand, if sources of CO_2 in addition to fossil fuel are present, which more than offset any such sinks, an even larger shortfall in oceanic uptake is apparent. The growing strength of this latter possibility (see Chapter 3) has encouraged the development of more realistic ocean models in an attempt to discover whether a better treatment of oceanic mixing can lead to greater uptake of CO_2.

From the discussion of oceanic mixing given earlier it seems unlikely that the composite effect of all mixing processes could be described by a constant, globally averaged diffusivity. The emphasis has thus been on developing models which include the main short-term mixing processes which operate on timescales of about 30 years, the mean age of fossil fuel CO_2. These are the ventilation of the thermocline and intermediate waters (<1000m) and the formation of new deep water in polar regions, illustrated schematically in Fig. 11.

Crane (1982) and Siegenthaler (1983) extended the box-diffusion model of Oeschger et al., to include, in different ways, direct exchange between the atmosphere and the deep water outcrops in high latitudes. They obtained airborne fractions somewhat reduced but still in excess of 0.60.

Both Broecker et al. (1980a) and Viecelli et al. (1981) introduced greater meridional resolution by constructing models in which parallel sections, characterised by vertical advection and diffusion, represented the equatorial, temperate and polar oceanic regions. The models were calibrated using the GEOSECS measurements of bomb-produced ^{14}C

and ^3H, which depict the short-term mixing processes, rather than natural ^{14}C whose distribution represents the effects of global mixing on timescales of several thousand years. Vertical eddy diffusion coefficients several times larger than deduced from natural ^{14}C profiles were obtained for temperate and polar waters. When extrapolated from the oceanic regions for which they were calibrated to the global ocean, these models tentatively suggest that the oceanic uptake may be larger, by as much as a factor of 1.5, than predicted by the one-dimensional, global box-diffusion model calibrated against natural ^{14}C profiles.

An alternative approach to the inclusion of intermediate and deep water formation was adopted by Bjorkstrom (1979) using a 'multi-box' model comprising a network of well-mixed reservoirs. Fig. 12 illustrates the structure of the model although the details of the atmosphere-biosphere exchanges and the biological fluxes in the ocean have been omitted for clarity. The surface layer is divided between cold and warm reservoirs, the cold water exchanging with two intermediate layers and being injected into a succession of deep water layers. The return flow to balance the deep water input is achieved by upwelling through the intermediate layers and surface transport via the warm water box. Calibration of a complex model such as this poses problems. In his original paper Bjorkstrom carried out sensitivity analyses for wide ranges of parameter values yielding airborne fractions in the range 0.25-0.70. Recently, Bacastow and Bjorkstrom (1981) have concluded that with what are probably realistic water mass fluxes and the assumption of a neutral biosphere this model, like the simpler box models before it, also predicts airborne fractions in excess of those observed.

Future progress in modelling oceanic uptake of CO_2 will depend to a large extent on the acquisition of data for calibration purposes and improved knowledge of ocean circulation rates. The National Science Foundation's Transient Tracers in the Oceans (TTO) programme, now in progress, has been designed specifically to look at these problems in relation to the carbon dioxide issue.

Having dealt with the uncertainties over the results from ocean uptake models arising from lack of knowledge of the relevant parameters, we now pay attention to effects not usually incorporated into such models. Most of the models used so far deal only with uptake of CO_2 by seawater, its reaction with other dissolved species and the rate at which all of these various aqueous forms are mixed into the main body of the oceans. Far less attention has been paid to interaction of the dissolved species with biota in the seawater and solid phases in bottom sediments or suspended in the water column. These two possibilities are considered next.

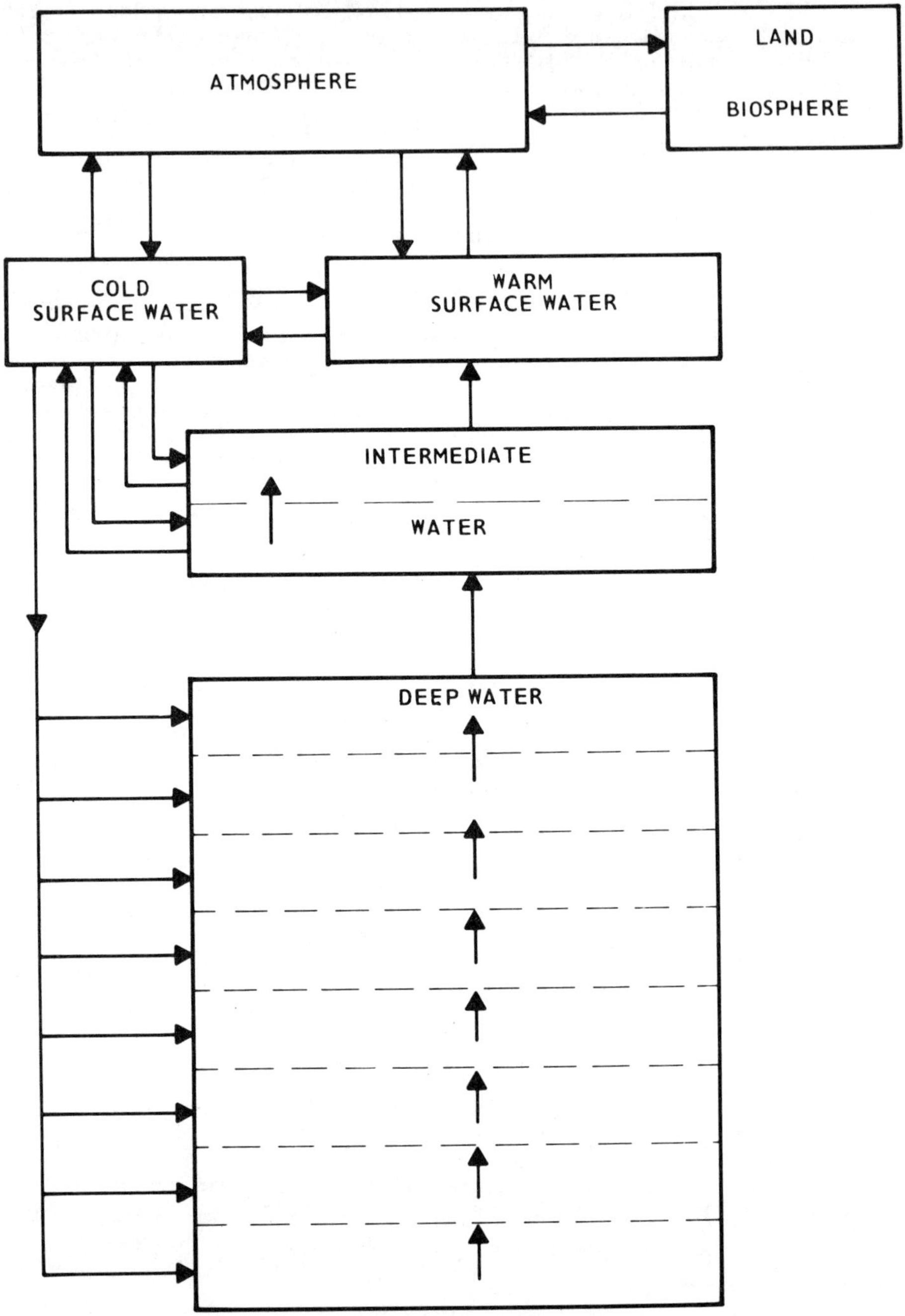

Figure 12. The multi-box model of Bjorkstrom (1979). Details of the atmosphere-biosphere exchanges and the biological fluxes in the ocean have been omitted for clarity.

Heterogeneous processes which might increase oceanic uptake of CO_2

a) Biological uptake

Marine plants are thought to fix about 25 GtC each year, of which about 95% is attributable to planktonic processes with 5% due to benthic and estuarine ecosystems. However, the figure of 25 GtC $year^{-1}$ represents gross fixation since recycling is rapid, on the timescale of weeks, in the euphotic zone. The quantity of fixed carbon escaping from the upper mixed layer into deeper water is of the order of 10% of the gross fixation rate and after oxidation at depth the amount sedimented out of the system into bottom deposits is probably only about 10% of this (i.e. ~1% of the gross carbon fixed, 0.25 GtC $year^{-1}$).

This rapid recycling in the surface layers and slow loss of detrital material to the sediments is part of the natural circulation of carbon by the marine biosphere. Thus, in order to help balance the budget for anthropogenically produced CO_2, the flux of detrital carbon to bottom sediments would need to increase. However, since there seems to be no possibility that increase in inorganic carbon content will by itself enhance marine production, any effect has to be indirect.

One possibility is that since the productivity of marine plankton appears to be nutrient rather than CO_2 limited, if man's activities have increased the input of plant nutrient elements (particularly phosphorus and nitrogen) into the oceans then the resulting enhanced production and subsequent sedimentation might contribute a valid route for uptake of additional CO_2 by the marine environment.

The first thing which must be said is that this idea will work, in the direction of increasing CO_2 uptake, only for fixation of carbon in an organic plant matrix; any nutrient-induced increase in organisms secreting calcareous shells would work in the opposite sense, as the following equation shows:

$$Ca^{2+}_{(aq)} + 2HCO_3^-{}_{(aq)} \rightleftharpoons CaCO_{3(s)} + CO_{2(g)} + H_2O_{(aq)} \quad \ldots(12)$$

The forward reaction represents the formation of a calcium carbonate solid phase and is accompanied by release of gaseous carbon dioxide. Only phytoplankton producing carbohydrate (or other forms of organic carbon) according to the classical photosynthetic reaction (equation (13)) constitute a potential sink for excess CO_2,

$$6\ CO_{2(g)} + 6\ H_2O_{(l)} \xrightarrow[\text{nutrients, chlorophyll}]{h\nu} \underset{\text{carbohydrate}}{C_6H_{12}O_{6(s)}} + 6O_{2(g)} \quad \ldots(13)$$

Several authors have attempted to calculate the additional carbon which might be sequestered in the oceans by this route. Although there is no doubt that man's activities have released additional nutrients to marine waters, e.g. sewage and fertilisers provide both soluble phosphorus and nitrogen, and detergents contain polyphosphates which break down in receiving waters to give phosphates, the difficulty comes in estimating not only the size of the augmented inputs, but also how much extra organic carbon is formed and sedimented out of the water column.

Mackenzie (1981) has approached the calculation by estimating the extra amounts of phosphorus and nitrogen flowing into the ocean each year as a result of man's activities. By combining these with a knowledge of the ratios of carbon to phosphorus and carbon to nitrogen in organic matter sedimented in the marine environment (150:1 and 10:1, respectively) he calculates that fertilization with phosphorus and nitrogen could lead to fixation of 0.048 (1%) and 0.108 (2%) GtC $year^{-1}$, respectively. The figures in brackets represent the approximate percentage of the amount of fossil fuel CO_2 presently entering the atmosphere which could be accounted for by this removal route. The figures should probably be treated as maximum values since it is assumed that all the extra nutrient leads to an equivalent amount of carbon sedimented out of the marine system. However, as Walsh et al. (1981) point out, this assumption may not introduce too great an error since there is evidence that in near-shore areas (where most of the increased production will occur) something like 50% of the carbon fixed ends up buried in the sediments. In their own attempt to calculate the extra carbon which may be sequestered by enrichment of coastal waters with nitrogen, Walsh et al. appear to obtain a figure of 0.76 GtC $year^{-1}$, which is about 7-8 times greater than the estimate of Mackenzie. The discrepancy arises in part because Walsh et al. use a larger anthropogenic inflow of nitrogen than that in Mackenzie's calculation, and also assume that each extra nitrogen atom is recycled several times per year.

Although the figure derived by Walsh et al. (1981) is large enough to account for the missing carbon in fossil fuel CO_2 budgets having a 'neutral' biosphere, it must only be accepted with great caution. The reasoning of Walsh et al. is sometimes unclear and the calculation involves several untested assumptions. Because of this, in the carbon budget to be presented in Chapter 4, only the lower figure calculated by Mackenzie (1981) will be used.

b) Reaction with marine minerals

Another way in which the oceans might take up more CO_2 is by reaction with appropriate mineral phases either in bottom sediments or suspended in the water column. The reverse of equation (12), in which water containing dissolved CO_2 reacts with a solid calcium carbonate to produce calcium and bicarbonate ions, is an example of how such reaction could occur. This is, of course, the same reaction which takes place on land when water containing CO_2 reacts with $CaCO_3$ in the weathering of calcitic minerals, a reaction which ultimately provides some of the natural alkalinity of seawater by inflow of $Ca^{2+}{}_{(aq)}$, $HCO_3^-{}_{(aq)}$ and $CO_3^{2-}{}_{(aq)}$ in rivers.

Although carbonate minerals occur in large amounts in oceanic sediments underlying shallow and mid-depth waters, the CO_2 sink represented by the reverse of equation (12) has generally been discounted, at least until recently. The reason for this is that if one calculates whether seawater is super- or under-saturated with respect to the two principal forms of pure calcium carbonate (calcite and aragonite) it turns out that surface waters are generally substantially supersaturated. This means that a large amount of CO_2 would need to be added to back off the supersaturation (via conversion of $CO_3^{2-}{}_{(aq)}$ to $HCO_3^-{}_{(aq)}$ as discussed earlier) before any dissolution of $CaCO_3$ could take place by the reverse of equation (12). In deep waters, which are often undersaturated with respect of $CaCO_{3(s)}$, the process is not important since, as might be expected, very little calcium carbonate solid phase is found in the underlying sediments.

The situation has changed somewhat recently with the realisation that in marine sediments the calcium carbonate mineral phases are often not pure calcite or aragonite. The calcitic minerals often contain considerable amounts of magnesium (up to 15%) which makes tham significantly more soluble in seawater. This greater solubility means that the water, rather than being supersaturated, may be close to equilibrium with the magnesium calcite mineral. If this is so then addition of CO_2 will lead to the possibility of direct reaction with the solid phase, for example (Wollast et al., 1980):

$$Ca_{0.88}\,Mg_{0.12}\,CO_{3(s)} + CO_2 + H_2O \rightarrow 0.88\,Ca^{2+}{}_{(aq)} + 0.12\,Mg^{2+}_{(aq)} + 2HCO^-_{3(aq)} \quad ...(14)$$

Such reaction is most likely to occur in nearshore and coastal areas due to both the presence there of high magnesium calcites and the shallow water depth, which presents little barrier to contact between the added CO_2 and the bottom sediments. Support for these ideas comes from the results of *in situ* dissolution experiments conducted in a shallow subtropical marine environment by Balzer and Wefer (1981).

Calculation of the size of this potential sink for excess CO_2 is extremely difficult, since it is necessary to estimate not only the amount of suitable sediment available for reaction (both its area and content of magnesium, since solubility depends on amount of magnesium in the mineral lattice), but also the kinetics of the dissolution reaction. The task was attempted at a Workshop organised by the US Department of Energy (US DoE, 1981) and the best estimate was that such mineral reactions might provide a sink for very approximately 0.07 GtC $year^{-1}$, i.e. about 1.5% of the amount currently produced by fossil fuel combustion.

Experimental approaches to estimating oceanic uptake

In view of the various uncertainties, reviewed earlier in the chapter, concerning quantitative prediction of the ability of the oceans to take up anthropogenic CO_2, it is pertinent to ask what possibilities exist for estimating the ocean sink by direct measurement.

The most obvious approach is to make secular measurements of the total dissolved inorganic carbon in the oceans. However, the change to be expected from such direct observations (about 0.035% year^{-1}) is too small, by a factor of about 10, to be detectable with presently available analytical techniques a variant of this idea, in which the total dissolved inorganic carbon in new, surface water is compared with that in older, deep water, has been attempted (Brewer, 1978; Chen and Millero, 1979) and the results are in broad agreement with those from ocean uptake models. However, although this is reassuring, it should be borne in mind that the method of calculation involves a large number of uncertainties; essentially how to correct for changes in the CO_2 content of the water since its formation at the surface, due to biological activity, reaction with carbonate solid phases, and mixing with other water masses Approaches of this type, which at present either lack or only just possess the necesary sensitivity, will become increasingly important in the future as the size of the 'signal' gets larger and techniques of chemical analysis improve.

Another potentially valuable approach is to make simultaneous measurements of the partial pressure of CO_2 in both air and surface seawater. From the sign and magnitude of the observed gradient and a knowledge of the transfer velocity for the air-water exchange, it is possible to calculate the next flux of CO_2 into or out of the seawater at the place and time the observations were obtained. If a temporally and spatially representative number of such observations can be made then the resultant global net flux might be computed. Since the size and sign of the air-sea CO_2 concentration difference is known to be highly variable both geographically and on various timescales (diurnal, seasonal), clearly a very large number of observations would be required, as well as knowledge of how to estimate the CO_2 transfer velocity for each set of measurements (see page 19 for a discussion of this last point).

A third possibility is to try to obtain the CO_2 flux across the sea surface by making measurements of the gas in the marine atmosphere using either the flux gradient or eddy correlation techniques. In the former, CO_2 concentrations in the air measured at two or more heights above the sea, and from a knowledge of the transfer rate through the atmospheric boundary layer the flux of CO_2 across the air-sea interface can be calculated. Although successfully employed for gases, such as SO_2, where the main resistance to transfer is on the air side of the interface, this technique is unlikely to be sufficiently sensitive for gases like CO_2, where the main transfer resistance (and hence most of the concentration gradient) is in the water phase. In the eddy correlation approach fast response sensors (response time <1 s) measure CO_2 concentration and the vertical component of the wind and, by summation of the products of these two quantities, the net flux into or out of the sea is obtained. This technique has been used on a trial basis (Jones and Smith, 1977; Wesely et al., 1982) but, as discussed previously in connection with the partial pressure difference approach, a very large number of measurements would have to be made before any meaningful estimation of the global air-sea flux of CO_2 could be made. However, recent objections to the use of the eddy correlation approach to obtaining CO_2 fluxes may mean that the potential utility of the technique cannot be realised (Bakan, 1978; Webb et al., 1980).

3 The Terrestrial Biosphere and Atmospheric Carbon Dioxide

As discussed in Chapter 1, seasonal variations in atmospheric CO_2 concentrations are generally ascribed to the photosynthesis-respiration/decomposition activity of the terrestrial biosphere. Now we turn our attention to the possibility that land vegetation may also play a role in longer-term changes in atmospheric CO_2 levels.

There are two ways in which this could occur. First, since plants abstract CO_2 from the air and convert it into fixed carbon (carbohydrate, etc. - see equation (13)) in the process of photosynthesis, will anthropogenically induced increased levels of atmospheric CO_2 lead to an increase in photosynthetic fixation of carbon dioxide? Long-term storage of this additional plant material in trees and soils could then constitute a valid sink for man-mobilized excess atmospheric CO_2. In contrast, the second possibility is that by the cutting of virgin forests and other changes to existing vegetation and soils man may have created an additional source of atmospheric CO_2. This would operate by the processes of respiration/decomposition and burning mobilizing CO_2 formerly sequestered in the biosphere as fixed carbon. By reference to equation (13), the first process can be envisaged as a shift of the equilibrium position from left to right, whereas the second process moves the equilibrium in the opposite direction, i.e. the two effects work in opposite senses with respect to their roles in altering levels of CO_2 in the atmosphere. They are discussed separately in the next two sections.

The possible effect of elevated atmospheric CO_2 on terrestrial photosynthesis

It is well established that in some circumstances addition of carbon dioxide gas can lead to enhanced growth of plants. For example, commercial producers of such crops as cucumber, lettuce and tomatoes under glass often add CO_2 to their greenhouses, which is found to lead to an increase in yield. However, although often cited as evidence for the ability of plants to extract excess CO_2 from air, the results from such artificial situations should only be extrapolated to the natural environment with great caution. Goudriaan and Ajtay (1979) have pointed out that without addition of CO_2, plants grown under glass may be starved of the gas since, with the restricted exchange with the outside, levels of CO_2 in greenhouses can drop to as low as 100 ppm. Furthermore, Lemon (1977) cites Dutch experience which indicates that although crops grown in greenhouses to which CO_2 has been added produce earlier and larger crops of lettuces, there is no significant increase in carbon fixed on a dry weight basis. All that appears to happen is that the plants grown in a CO_2-enriched atmosphere contain more water- which is then sold to the customer.

Probably of more relevance to what happens under natural conditions are the results of a large number of controlled laboratory and a smaller number of field studies in which plants have been grown under different partial pressures of CO_2 (work cited and described in Lemon, 1977; Loomis, 1979; Goudriaan and Ajtay, 1979; Downton et al., 1980; Loucks, 1980; Kramer, 1981; Rosenberg, 1981). The results of these experiments are in general agreement in showing that for C_3 plants (most of the temperate zone species) there is a continuous increase in rate of photosynthesis with increase in the partial pressure of CO_2 up to about 1000 ppm, above which point saturation occurs. In contrast, the situation for C_4 plants (many of the tropical grasses) is that CO_2 increase above ambient leads to no significant increase in growth. For these species the carbon dioxide saturation point appears to be at a partial pressure in the region of 200 ppm, i.e. well below the present atmospheric concentration.

The essential reason for this difference in behaviour between the C_3 and C_4 plants arises because the former photorespirate (biological production of CO_2 from O_2 under the action of light) and photosynthesise at the same time, whereas only the latter process occurs in C_4 species. Photorespiration leads to an increase in concentration of intercellular CO_2, via the production of the gas within leaf cells, which reduces the concentration gradient driving CO_2 across the leaf surface from the air; relative to the situation in C_4 plants where there is no inter-cellular production of carbon dioxide.

The potential increase in plant yield due to elevated partial pressure of CO_2 is often expressed in terms of a 'β-factor', which is defined as the percentage increase in photosynthesis resulting from a one per cent increase in atmospheric CO_2. Bjorkstrom (1979) cites a value for β of 0.2 or less. This is somewhat lower than the value which can be derived from the recent detailed treatment by Monteith (1981) who predicts the increase in plant yield to be 0.1% per 1 ppm CO_2 increase, corresponding to a β-factor of approximately 0.3.

The results discussed so far are all for plants grown under rather special situations (e.g. greenhouses, idealized laboratory conditions, or well regulated agricultural systems). In these artificial systems factors other than CO_2 which affect plant growth, e.g. availability of nutrients and water, are optimized as far as is possible. In contrast, in most natural and many agricultural situations this is not the case and shortages of nutrient elements and/or water are more limiting then carbon dioxide. Thus, it would appear that direct fertilization of plants by added CO_2 is unlikely in most natural and agricultural environments (Goudriaan and Ajtay, 1979).

Another possible way in which elevated levels of CO_2 may have an indirect effect on the growth of plants is by increasing the efficiency with which they use water. This comes about because when grown with elevated levels of CO_2 both C_3 and C_4 plants respond by partial closure of their stomata. This does not much affect the passage of CO_2 into the leaves, for which the rate controlling step is transfer across the walls and internal fluid of mesophyll cells (Monteith, 1981; Rosenberg, 1981), but does substantially decrease the rate at which the plant leaves lose water, which in most circumstances is controlled by the degree of opening of the stomata. No quantitative estimate of the possible importance of this effect in the environment seems to have been attempted, although it is clearly of potential importance since it may allow colonization of areas previously too dry for plant growth (Downton et al., 1980).

Although it seems reasonable to conclude, on the basis of present knowledge, that direct effects of increase in atmospheric CO_2 on plant growth are either unproven or unquantified, several authors (e.g. Loomis, 1979; Monteith, 1981) have stressed that changes in climate (whether resulting from CO_2 increase or from other causes) are likely to be of more importance than any direct result of elevated levels of CO_2. For example, changes in temperature or rainfall produce well documented and rather substantial effects on the growth of plants. Furthermore, elevated temperature will lead to an increase in respiration and decomposition which will speed the rate at which CO_2 is returned to the atmosphere by the biosphere and so tend to negate any increase in plant uptake as a result of higher levels of CO_2 in the atmosphere.

The possible effect of change in land use on atmospheric CO_2

Although it is obvious that man has altered the terrestrial biosphere in recent centuries via his agricultural and industrial activities and urbanization, it is only in the last five years that any attempt has been made at quantitative assessment of the possible effect of these changes on atmospheric carbon dioxide. Since industrial and urban developments clearly reduce the amount of vegetated land and agricultural crops often represent less fixed carbon per unit area than the forests and other cover they have replaced, it is generally reckoned that such changes in land-use result in a net input of CO_2 into the atmosphere. Furthermore, the size of the terrestrial reservoir of carbon (living biota plus soil and humus) is substantially larger than the amount of carbon in the atmosphere (see Fig. 16), so that even a small percentage change in the amount of organic carbon stored on land would be amplified considerably when converted into CO_2 increase in the atmosphere.

Probably the first estimates of the amount of CO_2 which might be added to the atmosphere from changing land use were those of Woodwell and Houghton (1977) and Woodwell et al. (1978). By using a figure for the present rate of forest clearing of 1-2% per annum, derived from studies by L.S. Hamilton for the forests in Venezuela, they calculated that

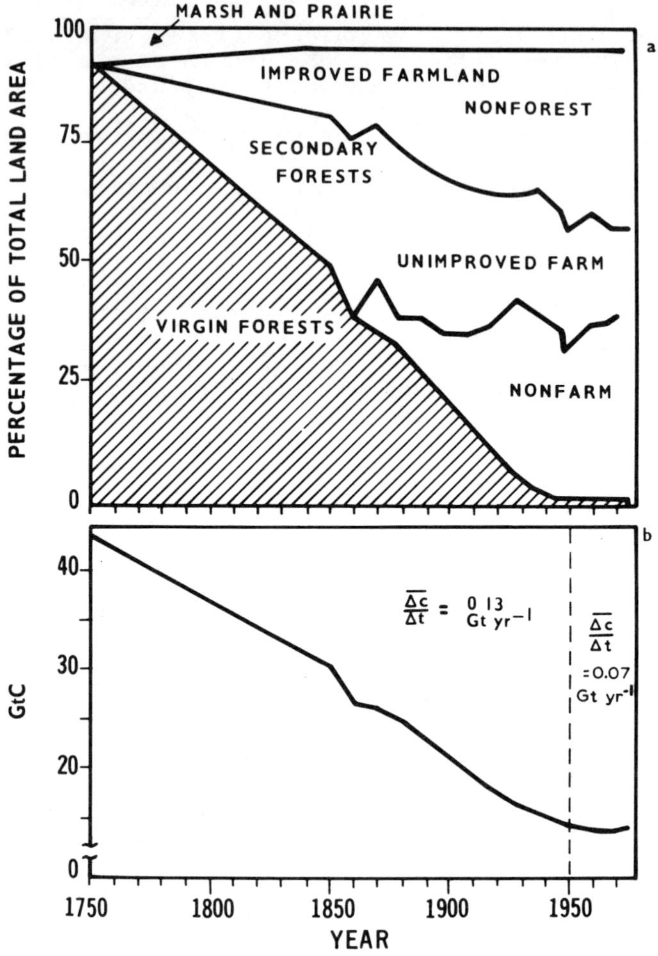

Figure 13. Land-use history (a), and changes in total carbon (b), for the south-eastern United States, 1750-1977. (After Delcourt and Harris, 1980, copyright 1980, the American Association for the Advancement of Science)

the most probable range for global CO_2 release resulting from change in the biosphere is 4-8 GtC $year^{-1}$. These figures are comparable with the rate of production of carbon dioxide by combustion of fossil fuels (5 GtC $year^{-1}$) and, if correct, would clearly be a very important term in the global budget for anthropogenically produced CO_2 (see Chapter 4).

However, Loomis (1979) has questioned the above rate of clearing for Venezuelan forest, as well as its applicability worldwide, which was assumed by Woodwell and his co-workers. Ralston (1979) suggests that from F.A.O. forestry statistics

Table 2: Estimates of Annual Net Carbon Release from Terrestrial Ecosystems to the Atmosphere in or About 1970. (From Moore et al., 1981)

Authors	GtC year^{-1}
Woodwell et al. (1978)	4 to 8
Hampicke (1979)	1.5 to 4.5
Adams et al. (1977)	0.4 to 4
Bolin (1977)	0.4 to 1.6
Wong (1978)	1.9
Moore et al. (1981)	2.2 to 4.7

the rate of carbon removal from forests is about an order of magnitude less than the value used by Woodwell's group in their calculation. Broecker et al. (1979) stress the importance of including estimates of regrowth of plants on forestry land after clearance. Bolin (1977) using a similar approach to Woodwell's, but with more detail for temperate and boreal forests, cautiously concludes that the current net rate of release of CO_2 from the biosphere to the atmosphere is probably in the range 0.4-1.6 GtC year^{-1}, i.e. significantly less than the original estimates from Woodwell's group. Probably the largest uncertainty with Bolin's calculation arises from the paucity of representative data for vegetated tropical areas.

The above values for the net carbon exchange between terrestrial ecosystems and the atmosphere, as well as estimates by other workers, are given in Table 2. All the numbers given in the Table refer to the situation in or about 1970. It is pertinent to ask what the position was in the previous two hundred years, which covers the period of so called 'pioneer' settlement and conversion of virgin land to agricultural use. This has been attempted on a regional scale by Delcourt and Harris (1980) who have used agricultural and forestry statistics for the south-eastern USA to estimate net carbon transfer between this area and the atmosphere in the period 1750-1980. Their results are shown in Fig. 13 where the dramatic change from virgin forest to various other forms of land use is clearly shown in graph a. Delcourt and Harris calculate that right up until 1950 this area was a net source of CO_2 for the atmosphere, the average release rate over this 200-year period being 0.13 GtC year^{-1}. Since 1950 the increased productivity of commercial forests has reversed the direction of the net flux and over the last 30 years there has been net uptake of CO_2 from the atmosphere at an average rate of 0.07 GtC year^{-1}.

Using a modelling approach, Moore et al. (1981) have attempted to assess the direction of the net CO_2 flux between terrestrial ecosystems and the atmosphere on a global basis for the period 1860-1970. They divide the land surface up into 10 geographic regions each containing potentially 12 different ecosystems. Using data for the annual rates of change in land use and idealized models for the time course of the resulting uptake or release of CO_2, an integrated global picture of the annual net flux is obtained. An important finding from this study is that clearing of land for agriculture results in release of appreciably more CO_2 than clearance purely for forestry. In the latter case regrowth generally takes place soon after the land is cleared, whereas change to agricultural use often results in continued ploughing and release of CO_2 by breakdown of soil organic matter. The results of the global model of Moore et al. (1981) are shown in Fig. 14 together with the flux of carbon from fossil fuel combustion as estimated by Rotty, and the sum total of these two net transfers. According to the results of Moore et al., the globally integrated biospheric flux of CO_2 in the period examined was always from land to air. Up until about 1960 the flux from the terrestrial biosphere exceeded the input of CO_2 to the atmosphere from the burning of fossil fuel, although since then the reverse has become the case. The model predicts that in 1970 the net biospheric release of CO_2 was in the range 2.2 to 4.7 GtC $year^{-1}$ and these values are included in Table 2.

It should be stressed that the work of Moore et al. suffers from a number of uncertainties. There are difficulties with the input data (areas and types of vegetation and their change with time from 1860-1970), especially for tropical areas, and the models utilized for the time course of CO_2 exchange following alteration in land use are idealized and not subject to experimental verification. Furthermore, no allowance is made for possible fertilization of plant growth due to elevated atmospheric CO_2 levels (as discussed earlier in this chapter) which would tend to decrease the net flux from the biosphere. Another area of uncertainty is that in the model it is assumed that all biomass left in the land area being cleared will ultimately be converted to CO_2. This ignores the production of charcoal by biomass burning which Seiler and Crutzen (1980) estimate at 0.5-1.7 GtC $year^{-1}$. Since this material is very resistant to oxidation to CO_2, it may constitute a significant long-term sink for anthropogenic carbon dioxide (UNESCO, 1980).

Possible indicators of the net effect of the terrestrial biosphere on atmospheric CO_2

As discussed earlier, the two possible effects of man's activities on the biosphere (i.e. increase in biomass resulting from elevated atmospheric CO_2, and reduction in biomass due to changing land use) appear most likely to operate in opposite senses with respect to their effects on levels of atmospheric CO_2. Thus, to obtain the net result of both processes requires a quantitative knowledge of each of them. However, it is clear that present understanding is largely qualitative, and where quantitative estimates have been attempted they are subject to considerable uncertainty.

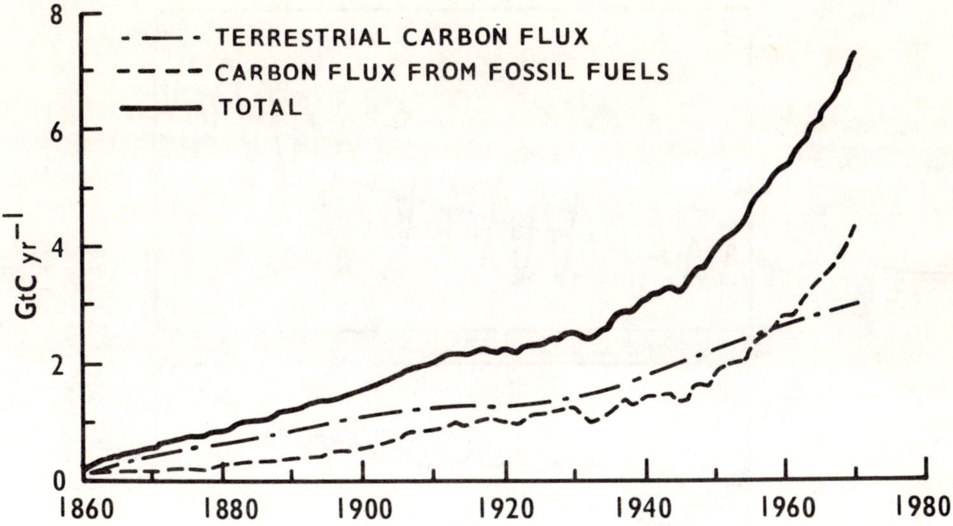

Figure 14. Annual net flux of carbon from the terrestrial biosphere, fossil fuel consumption and cement production (Keeling, 1973a; Rotty, 1973, 1974), and the sum of these two. (From Moore et al., 1981, reprinted with permission from J. Wiley and Sons)

In these circumstances a potentially valuable alternative approach is to find some environmental parameter which will indicate the net effect of the terrestrial biosphere on atmospheric CO_2. There is something to be said for such an approach since, apart from the obvious difficulties of direct estimation of the individual biospheric fluxes, for several aspects of the general problem, e.g. global budgeting of man-induced production of CO_2 and the role of elevated atmospheric levels of the gas on climate, only the net flux is required. There are two environmental parameters which have been suggested as possible indicators of net biospheric flux: the magnitude of the seasonal variation in atmospheric CO_2, and measurements of carbon isotopes in biological material. These are discussed in the following two sections.

a) Seasonal variation in atmospheric CO_2

The well marked seasonal variations in atmospheric CO_2 concentrations for Mauna Loa and other sampling stations were discussed in Chapter 1 - the decrease from spring to autumn and increase over the rest of the year being attributed to net photosynthetic consumption, and production due to net respiration/decomposition, respectively. Since this seasonal cycle seems to be largely attributable to the activity of the terrestrial biosphere, Hall et al. (1975) have used the Mauna Loa record of p_{CO_2} from 1958-1972 to try to see whether alterations in land biomass in this period have led to any change in the magnitude of the seasonal variation. The results of their study are shown in Fig. 15. Graph (a) shows

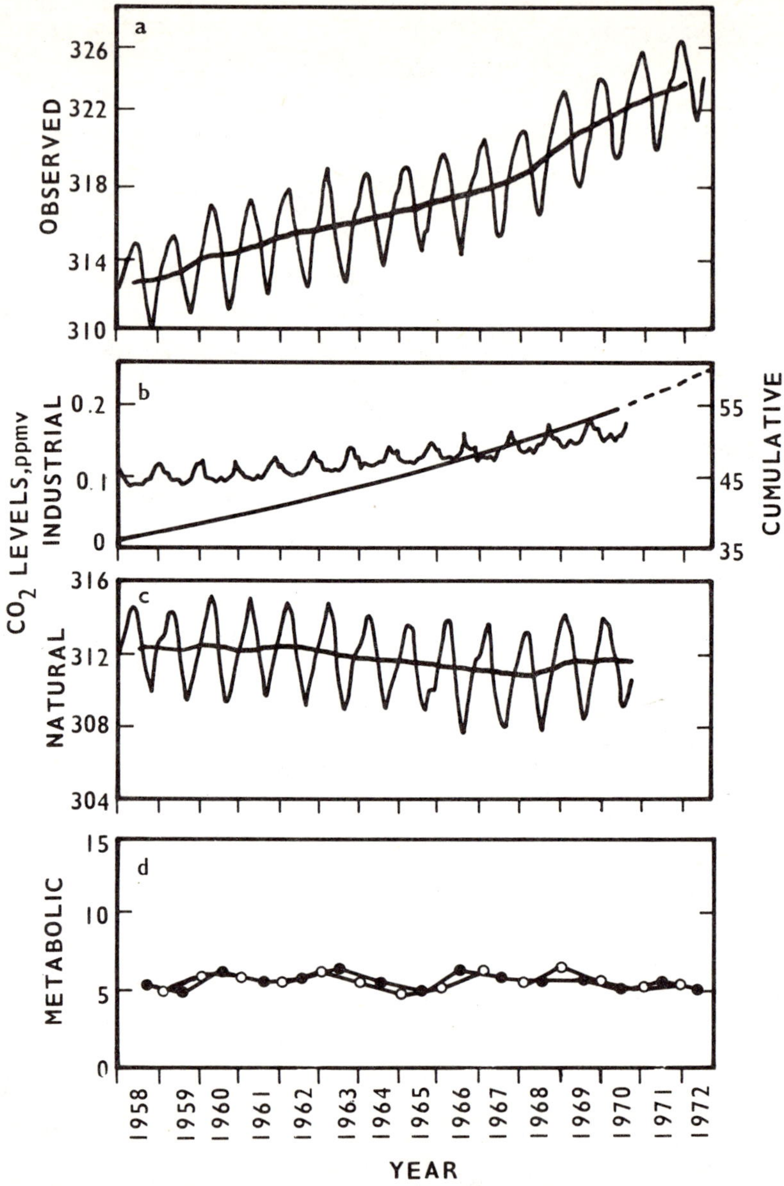

Figure 15. (a) Month-by-month variation in the atmospheric CO_2 concentration at Mauna Loa. The upward smooth curve is normalised for the semiannual fluctuations. (b) Month-by-month addition of industrial CO_2 to the Northern Hemisphere, along with the cumulative production above an assumed preindustrial level of 293 ppm. The last two years shown are extrapolated values from the least squares fit of the data to a third-order polynomial. (c) The observed CO_2 variation corrected for net industrial additions. (d) Semiannual net ecosystem production and respiration determined as the difference between crests and troughs in (c). Oceanic uptake of industrial additions taken as 42%. (From Hall et al., 1975, reprinted with permission from Nature, copyright 1975, Macmillan Journals Ltd.)

the observational record and this is adjusted for fossil fuel CO_2 emissions (graph (b)), assuming that 42% of the amount emitted is taken up by the oceans, to yield graph (c). For each cycle of curve (c) the partial pressure differences from peak to trough and trough to peak give, respectively, the net production and respiration of the terrestrial biosphere for that 12-month period. The values are plotted in graph (d) from which no clear trend is discernible.

Absence of any observable effect does not necessarily mean that significant changes have not occurred in the biosphere. It could be that decrease in biomass due to forest cutting etc., has been largely offset by increase brought about by fertilization of plant growth due to higher levels of atmospheric CO_2. Other uncertainties arise because of the shortness of the record used and the large size of the terrestrial biomass relative to man's ability to change it substantially. This last point may be quantitatively examined as follows. According to Table 1.1 of Bolin et al. (1979) the size of the (living) land biota ranges from 592 to 976 (mean about 800) GtC. Using Moore et al.'s (1981) estimate of the yearly net loss of terrestrial biosphere of 2.2-4.7 (mean 3.5) GtC year^{-1}, the total decrease in land plant matter over the 15 years (1958-72) is close to 50 GtC, or approximately 6%. A change in seasonal range of atmospheric CO_2 of this magnitude is probably near the sensitivity limit of the technique. However, in this calculation it has been assumed that all the decrease in biosphere comes from living material, whereas a substantial part probably comes from decay of soil and humus. Thus, the 6% change calculated above is a maximum figure, which implies that a greater length of record would be needed before the analysis could be expected to yield any meaningful trend. Recently, Pearman and Hyson (1981) have used a similar approach, but with the longer record available to them (1959-78), and conclude that there is now some evidence for a net increase in the size of the northern hemisphere biosphere over this period.

b) Isotopes of carbon

Carbon dioxide in the atmosphere is isotopically about 99% ^{12}C, with approximately 1% of the stable isotope ^{13}C and a much smaller amount of the radioactive ^{14}C. When C_3 plants (which constitute the dominant group in the biosphere) take up atmospheric CO_2 in photosynthesis they impose a fractionation on the $^{13}C/^{12}C$ ratio of about 17 per mil in favour of the lighter isotope. Thus changes in the size of the terrestrial ecosystem will lead to alterations in the $^{13}C/^{12}C$ ratio in the atmosphere. If biomass is destroyed and converted to CO_2 then the ratio of $^{13}C/^{12}C$ in the atmosphere will decrease and vice versa for net formation of new biomass. Hence the $^{13}C/^{12}C$ ratio in the atmosphere will depend on the size of the terrestrial biosphere and although photosynthesising plants will impose a fractionation, the $^{13}C/^{12}C$ in new plant material will reflect the atmospheric ratio at that time, *all other things being equal*. Thus, in principle, examination of changes in $^{13}C/^{12}C$ ratio in long-lived plants (e.g. trees) should give an indication of any alteration in the size of the terrestrial biosphere over the time period covered by the material examined.

Although simple in principle, as is often the case in environmental studies, there are substantial difficulties with this approach to estimating changes in size of the terrestrial biosphere, i.e. all other things are not equal and the inequalities are difficult to account for quantitatively. To begin with, burning of fossil fuel (which has a $^{13}C/^{12}C$ ratio similar to that in modern plants) also leads to changes in the atmospheric ratios of carbon isotopes. This significantly complicates estimation of changes in biomass size over the last 100 years, i.e. much of the period of interest in the present context. In order to make allowance for the fossil fuel-produced changes in $^{13}C/^{12}C$ use is made of $^{14}C/^{12}C$ ratios since, although modern plants and soils have a similar ratio of radioactive to stable carbon to that in the atmosphere, fossil fuel contains no ^{14}C (with a half-life of about 5700 years it has all long since decayed), and so CO_2 from this source dilutes the $^{14}C/^{12}C$ ratio in the atmosphere (the Suess Effect). By this means it is possible to separate out changes in $^{13}C/^{12}C$ ratio due to changes in the biosphere and fossil fuel consumption. One complication in this procedure is that the $^{14}C/^{12}C$ ratio in the atmosphere is subject to alteration as a result of changes in cosmic ray flux (by which ^{14}C is produced naturally in the upper atmosphere) as well as by input of 'dead' carbon from fossil fuel combustion. A more important problem is that nuclear weapons testing has greatly perturbed the levels of ^{14}C in the atmosphere in what is perhaps the period of greatest interest, i.e. the 1950s onwards when accurate direct observations of p_{CO_2} in the atmosphere became available, so that for this period the normalization procedure is not applicable (Broecker, 1980a).

As well as these problems due to fossil fuel combustion, other difficulties in deconvoluting the $^{13}C/^{12}C$ record in trees arise because the amount of ^{13}C taken up by plants varies with temperature and time in the growing season when the CO_2 is taken up. Furthermore, interpretation of the $^{13}C/^{12}C$ observations in terms of release of CO_2 by the biosphere depends on assumptions made about the net uptake of carbon dioxide by the oceans and biosphere and the fractionation occurring during these exchanges. Several of these problems can be approached by the isotopic analysis of sufficient trees having different temperature histories and with as wide a geographical coverage as possible. Unfortunately, to date only very few analyses of ^{13}C are available.

Notwithstanding all these difficulties, several authors have attempted to use measurements of $^{13}C/^{12}C$ ratios in trees to assess the net release of CO_2 by the biosphere over the last century. Their results are given in Table 3 and seem to indicate a net release of between 70 and 200 GtC over this period. Recently, the apparent concensus indicated by the results in Table 3 has been brought into question by the findings of Francey (1981) who has analysed $^{13}C/^{12}C$ trends over the last 200 years in two tree species in Tasmania and finds no real change with time. Since the fluxes given in Table 3 rely on N. Hemisphere observations, at a minimum the results of Francey imply that the results in the Table are not applicable on a global basis, and may mean that a more thorough reappraisal of all the data is required.

Table 3: Estimates of the Net Release of Carbon from Terrestrial Biota and Humus to the Atmosphere During the Last Century Based on Models Using Carbon Isotope Ratios in Tree Rings (from Moore et al., 1981)

Author	Interval	Release of carbon from terrestrial biota and humus to atmosphere during interval (Gt)
Stuiver (1978)	1850-1950	120
Wagener (1978)	1800-1935	170
Freyer (1978)	1860-1974	70
Siegenthaler et al. (1978)	1860-1974	133-195
Tans (1978)	1850-1950	150

Most authors seem to agree that there has been little change in $^{13}C/^{12}C$ ratio (corrected for the Suess Effect) since 1940 implying that any net biospheric release must have been before that date.

Conclusions on the role of the biosphere in the anthropogenic CO_2 cycle

Our overall impression is that man's role in altering the part which the terrestrial biosphere plays in the global cycling of CO_2 is the least well established element in the whole cycle. For almost every aspect discussed in this chapter the evidence is either very sketchy and/or equivocal in the conclusions which can be drawn from it. We probably cannot do better than quote from the conclusions of the WMO group of experts which reported recently (WMO/ICSU/UNEP 1981):

> 'In conclusion, it seems appropriate to adopt a value for net atmospheric emissions from the terrestrial biosphere since early in the last century between 75 and 175 GtC and an annual release at present between 0 and 4 Gt. The lower parts of these ranges are more likely to be applicable'.

The airborne fraction

In Chapter 1 the airborne fraction is defined as the ratio of the observed atmospheric increase in a particular period to the total CO_2 released from fossil fuel combustion and cement production during that period. This definition only accurately describes the partitioning of CO_2 between the atmosphere and other sinks when fossil fuels and cement are the only sources. If some part of the change in atmospheric CO_2 is due to biospheric effects then the partitioning implied by the calculated airborne fraction will be in error. To deal with this Oeschger and Heimann (1981) have attempted to calculate the proportion of the observed atmospheric CO_2

increase in the period 1957-78 which is due to fossil fuel combustion and cement production over the same period. The ratio of this corrected increase to the emission of CO_2 from fossil fuels and cement they term the 'theoretical' airborne fraction (the uncorrected ratio being called the 'apparent' airborne fraction). They point out that while net release of CO_2 from the biosphere would lead to the apparent airborne fraction over-estimating the amount of fossil fuel CO_2 remaining in the atmosphere, the decaying residual of a 'pioneer' release prior to 1957 would result in the apparent airborne fraction being an underestimate of the real atmospheric retention. Oeschger and Heimann calculate that for a constant net biospheric release equivalent to 0.5 ppm year^{-1} since 1860 the theoretical airborne fraction could be as low as 0.38, whereas if a total biospheric release (equivalent to a partial pressure of 40 ppm) was centred around 1890 the theoretical airborne fraction could be as high as 0.67.

It should be noted that these figures probably represent the extreme range of the true airborne fraction. It is likely that during the period of accurate observations of atmospheric CO_2 the decay of a pioneer effect and a net biospheric release have both been operating. Because of the uncertainty in our knowledge of the behaviour of the biosphere both now and historically, it is not possible to say which of these processes is or has been dominant.

4 The Global Budget of Man-Made Carbon Dioxide

The results of Chapters 1-3 are brought together in Fig. 16, which is an attempt to show quantitatively how CO_2 mobilized by man can be budgeted globally. Also shown are the sizes of the various environmental reservoirs for carbon, as well as the natural fluxes of CO_2 between them.

Although the sizes of the present-day natural fluxes (shown by arrows in Fig. 16) are somewhat uncertain, they are generally assumed to be balanced on a yearly basis, i.e. time and space averaged values of the air-sea and air-terrestrial biosphere fluxes are zero. In order to distinguish them from these natural transfers, the anthropogenic fluxes are encircled in the figure. Of the man-induced fluxes the best established is the input from fossil fuel burning (5 GtC year^{-1}) and of this about 2.5 GtC year^{-1} appears to remain in the atmosphere to produce the well established secular increase in CO_2 concentration. Best estimates of the amount which is taken up by the oceans (~2 GtC year^{-1}) are close to being able to balance the anthropogenic cycle, assuming no net exchange with the terrestrial biosphere. However, as soon as a significant net input from the land is included the budget will not balance unless additional sinks can be identified. Two such are shown in Fig. 16 (reaction with marine minerals and fertilization of near-shore waters) but they do not appear to be of sufficient magnitude to account for anything but very small amounts of CO_2. As discussed previously, both the magnitude of the net terrestrial input (and maybe even its direction) and the size of the additional marine sinks are very poorly established, as indicated by the query marks shown against these fluxes in Fig. 16.

One possible way of rationalizing the anthropogenic budget is to assume that in recent years the biosphere has been, at most, only a small net source of CO_2 to the atmosphere but was a larger one in the past (Stuiver, 1978; Delcourt and Harris, 1980). Then the present budget could be at or near to balance and, by assuming an atmospheric CO_2 partial pressure before major industrialization of less than the currently accepted 290 ppm (for example 260-270 ppm as suggested by Bjorkstrom, 1979; Stuiver, 1978; and Wigley, 1983), the atmospheric sink for CO_2 is substantially increased. Although this scenario is attractive, it is heavily

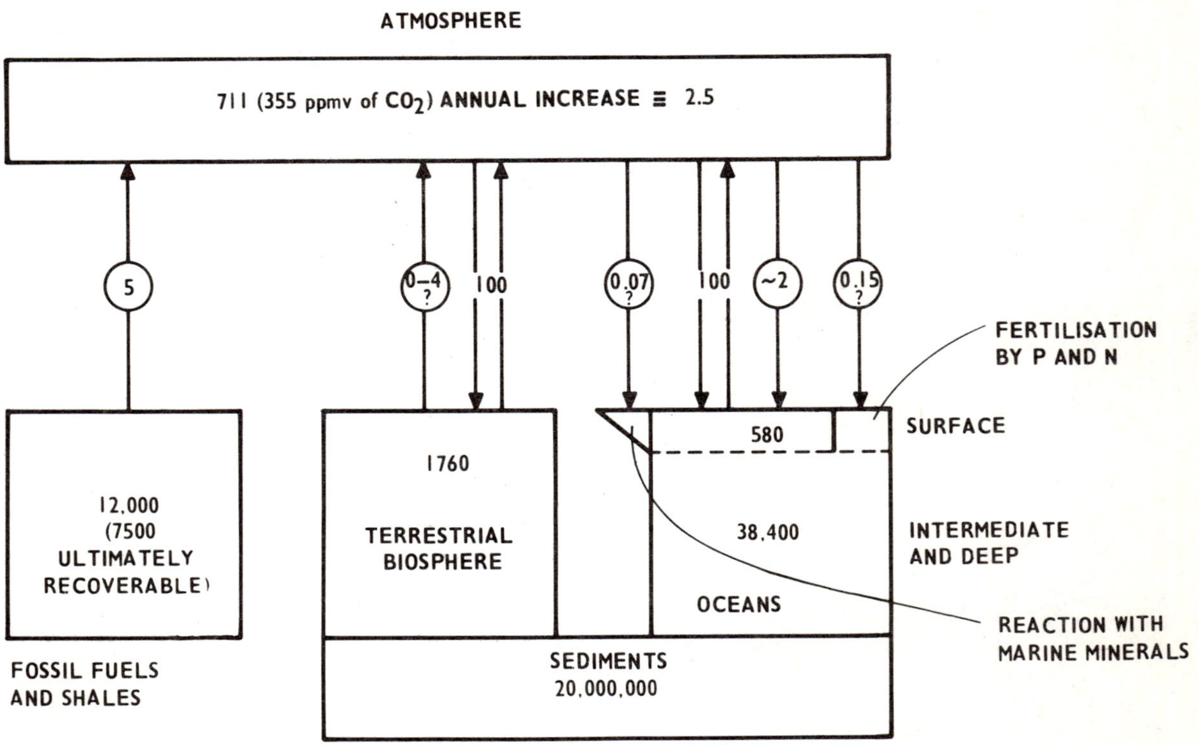

Figure 16. Global carbon reservoirs and present natural and anthropogenic fluxes between reservoirs. Reservoir sizes in GtC. Fluxes between reservoirs in GtC $year^{-1}$. Anthropogenic fluxes are circled.

dependent on a knowledge of the time course of any net biospheric release of CO_2 which, as we have seen, is far from well established.

With such uncertainty over the budgeting of anthropogenic CO_2, prediction of how future emissions will be partitioned between the various reservoirs and, especially in the climatic context, how much will remain in the atmosphere, is clearly difficult. The problem is often circumvented (but not overcome) by assuming for the proportion of fossil fuel CO_2 which appears to remain in the atmosphere a value close to 0.56 (the apparent airborne fraction calculated from the Mauna Loa record - see Table 1). With the present uncertainty over the budgeting of man-made CO_2 and the role of the biosphere, such an extrapolation is open to criticism, but is probably the best that can be done in the circumstances.

PART II

CLIMATIC CONSEQUENCES OF INCREASED CARBON DIOXIDE CONCENTRATIONS

5 The Greenhouse Effect

There is little doubt that in the absence of other climatic perturbations an increase in atmospheric carbon dioxide concentrations will give rise to globally averaged warming of the lower atmosphere. The degree of warming for a given increase in the carbon dioxide concentration is, however, difficult to predict and will depend on the complex interactions of physical processes in the atmosphere-ocean-cryosphere-lithosphere-biosphere system which control climate. The warming, which is likely to vary geographically, could ultimately be larger than any rise in temperature experienced in the past 10 000 years. Associated with the warming are likely to be corresponding changes in the distribution of rainfall and the patterns of the atmospheric circulation. The impact on society of climatic change occurring on this possibly very large scale, and over a period of time very short in climatological terms, could be considerable.

Although the possibility of a CO_2-induced climatic change was first studied quantitatively by Callendar (1938), much of the research on the likely magnitude of the change has been done in the last fifteen years with the aid of numerical models of climate. These simulate, with varying degrees of simplification, the response of the climate system to specified, larger carbon dioxide concentrations. The direct effects on the Earth's radiation budget of larger carbon dioxide concentrations (the 'greenhouse' effect) and the subsequent feedback processes triggered by the initial changes in the climate system must be modelled. However, even the most sophisticated models lack realistic representations of many of the physical and dynamical processes involved, and their simulations can, at present, only serve as a guide to the way in which the climate may change.

This chapter deals only with the direct radiative influence of increased carbon dioxide and shows how the radiative balances of the stratosphere, the troposphere and the surface are perturbed. It is intended to provide some background information for later chapters reviewing the variety of numerical models which have been used to simulate the eventual global climatic responses triggered by the initial radiative perturbations.

Atmospheric infrared absorption

Consideration of the global radiation balance shows that an effective temperature, T_e, for the Earth-atmosphere system can be obtained by equating the net incoming solar radiation at the top of the atmosphere with the outgoing terrestrial infrared radiation, with

$$\pi a^2(1 - \alpha)S = 4\pi a^2 \sigma T_e^4 \qquad \ldots (15)$$

where S is the normal solar flux incident at the top of the atmosphere, α is the planetary albedo, a is the Earth's radius and σ is the Stefan-Boltzmann constant. With α taking the observed value of around 0.3, T_e = 255 K.* The observed global mean surface temperature is close to 288 K, which indicates that the surface emits about 60% more radiation than is necessary to balance the incoming solar radiation at the top of the atmosphere. Since the constituents of the atmosphere which adsorb also emit infrared radiation they contribute to the outgoing flux at the top of the atmosphere, so the absorption of surface emission by the atmosphere is considerably more than the 60% excess. Since the atmospheric absorbers emit in all directions there is a downward infrared flux at the surface which contributes to the surface heating.

The chief absorbers of infrared radiation in the atmosphere are clouds, water vapour, carbon dioxide and ozone. Like the Earth's surface, most clouds absorb and emit almost as black bodies whereas the gaseous absorbers are active in well-defined spectral regions. Important are those gases which have absorption bands close to the peak of the Earth's emission spectrum. Fig. 17 shows the spectrum of blackbody emission at 272 K, approximating the Earth's emission at that temperature, and the downward longwave flux at the surface for a typical mid-latitude winter temperature profile. Of particular note in the figure are the vibration-rotation band of CO_2 around 15 μm, near the peak of the Earth's emission spectrum, the water vapour vibration-rotation band at 6.3 μm and the water vapour rotation band beyond 20 μm. Water vapour absorption also overlaps the longer wavelengths of the carbon dioxide absorption band centred at 15 μm. The coincidence of the Earth's emission spectrum and the atmospheric emission spectrum in the above wavelength regions indicates that all the radiation emitted by the surface is returned, none escaping to space. It is clear that water vapour is the most important infrared absorber in the atmosphere.

Fig. 18 further illustrates the influence of atmospheric absorption on the upward emission from the surface and the atmosphere by showing satellite measurements of the radiation emerging vertically from the top of the atmosphere. Superimposed are black-body emission spectra at various temperatures which provide a scale against which to estimate the

---

* Throughout Part II temperatures will be expressed in Kelvins (K) as is customary in climate modelling. 1K = 1°C; 273K = 0°C.

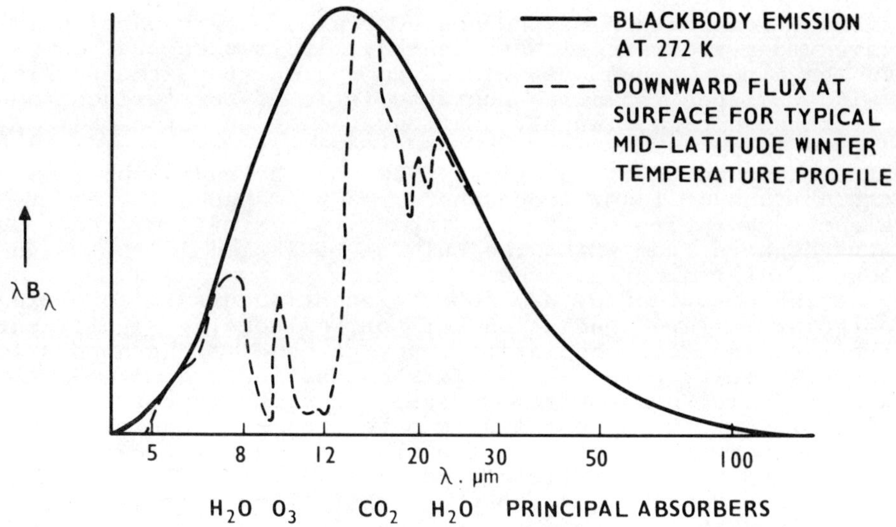

Figure 17. The Earth's black-body emission spectrum at 272 K and the downward flux at the surface for a typical mid-latitude winter temperature profile. The principal absorbers in different wavelength intervals are indicated. (Adapted from Roach and Slingo, 1979).

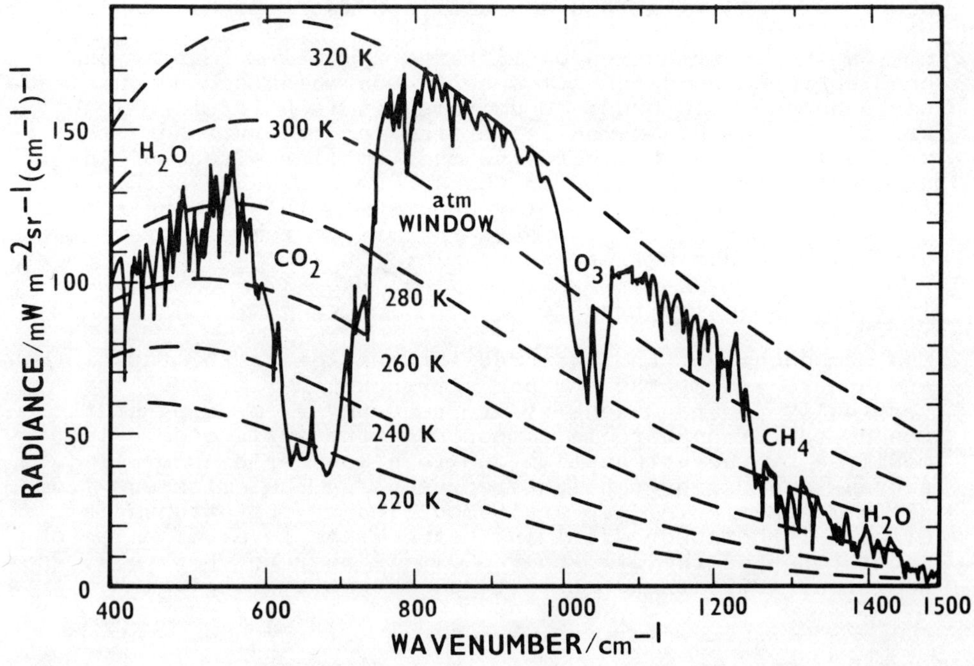

Figure 18. Thermal emission from the Earth plus atmosphere emitted vertically upwards as measured by the Michelson interferometer spectrometer on Nimbus 4 over the Sahara. The radiances of black-bodies at various temperatures are superimposed. (From Hanel et al., 1971).

average temperature of the source region at any particular wavenumber (reciprocal wavelength). It is seen that only in the almost transparent 10-12 μm region (the atmospheric 'window') does radiation characteristic of the surface temperatures emerge, largely unattenuated by the atmosphere. Around 15 μm (625-725 cm^{-1}) the atmospheric absorption is strong due to carbon dioxide. None of the radiation from the surface or lower troposphere reaches space, the equivalent temperature of 220 K being characteristic of the lower stratosphere from which the bulk of radiation in this wavelength interval originates.

The behaviour of atmospheric constituents in absorbing outgoing infrared radiation but not influencing significantly the incoming solar radiation (they have no strong absorption bands at solar wavelengths) is known as the greenhouse effect, analogy being drawn with the same radiative properties of glass which were once thought to be the major cause of warmth in greenhouses. The radiative transfer calculations of Manabe and Strickler (1964) indicate that if all the CO_2 were removed from the atmosphere, but other constituents maintained their present concentrations, the mean surface temperature would be some 10°C lower. Since lower temperatures would reduce the atmospheric concentration of water vapour and hence its contribution to the greenhouse effect, the actual reduction in temperature would be larger. This water vapour feedback effect is of crucial importance in evaluating the climatic consequences of an increase in atmospheric CO_2 and will be discussed in Chapter 6. It is first useful, however, to examine the direct influence of increased CO_2 on the distribution of infrared radiative heating and cooling throughout the atmosphere upon which subsequent feedbacks depend. In order to do this, it is helpful to look briefly at the basic radiative transfer theory. Note that we shall not consider here the much smaller effect of increased CO_2 on absorption in the 4.3 μm CO_2 band in the near infrared portion of the solar spectrum. This accounts for only about 13% of the increase in surface temperature according to Ramanathan and Coakley (1978).

Infrared radiative transfer

The components of the infrared intensities at the top of the atmosphere and at the surface are shown in Fig. 19. The vertically upward intensity at frequency ν emerging at the top of the atmosphere is composed of the surface emission, modified by the attenuation due to atmospheric absorption, and an integral over height representing the emissions from all atmospheric levels, again modified by the attenuating effect of the absorbers lying between the level of emission and the top of the atmosphere. It is easy to show (e.g. see Houghton, 1977) that the vertical intensity is given by

$$I_\nu = B_\nu(T_s)\tau_\nu^{z_s,\infty} + \int_{z_s}^{\infty} B_\nu(T(z))K_\nu(z)dz \quad ...(16)$$

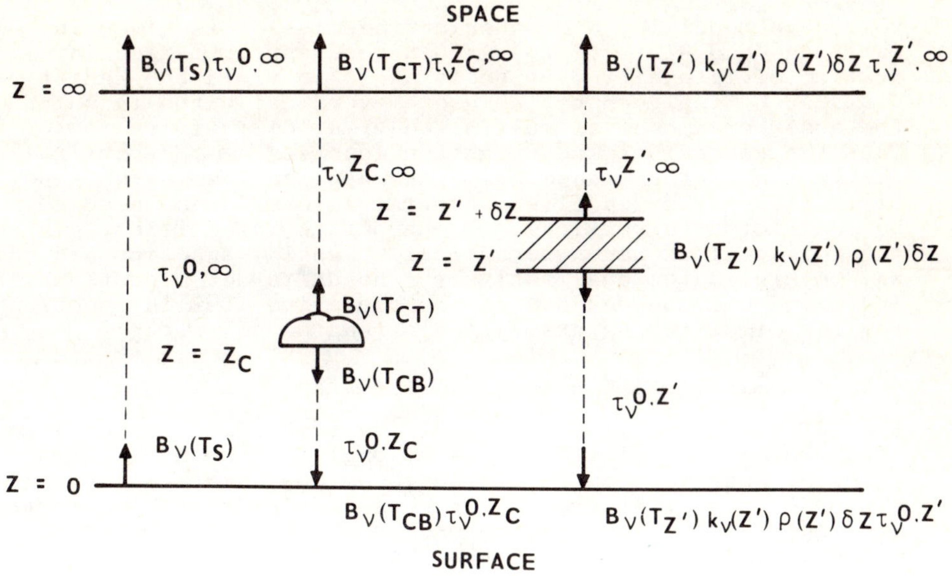

Figure 19. Components of the upward and downward infrared intensities at the top of the atmosphere ($z = \infty$) and the surface ($z = 0$) respectively. T_{CT}, T_{CB}, T_s are cloud-top, cloud-bottom and surface temperatures, k_ν is the absorption coefficient at frequency ν, ρ is absorber density; other symbols are explained in the text.

where $K_\nu(z) = d\tau_\nu^{z,\infty}/dz$, $\tau_\nu^{z,\infty}$ is the fractional transmission at spectral frequency ν of the atmospheric path from height z to space, subscript S refers to values at the Earth's surface (or a cloud top) and $B_\nu(T(z))$ is the temperature (T)-dependent Planck or blackbody function given by

$$B_\nu(T) = \frac{C_1\nu^3}{\exp(C_2\nu/T) - 1} \qquad \ldots (17)$$

where C_1 and C_2 are constants.

The atmospheric contribution to the intensity is seen to be a weighted integral of the Planck function profile, the weighting function, $K_\nu(z)$, being the derivative of the transmission to space. The downward intensity at the surface is given by a similar equation, the relevant transmissions being between height (z) and the surface, although there is no term equivalent to the surface contribution unless cloud is present. Integration of equation (16) over frequency and solid angle yields the total upward infrared flux.

Since the absorption coefficient, and thus transmission, is a very rapidly varying function of frequency, integration over frequency is extremely time-consuming if performed line by line. In climate modelling, methods are employed which enable average transmissions of finite intervals of the spectrum to be calculated. A discussion of the procedures involved in modelling atmospheric transmission is given in Paltridge and Platt (1976). To a good approximation integration of intensity over solid angle to yield the radiative flux can be replaced by the use of vertical paths in which the absorber amount is multiplied by a 'diffusivity factor'. Thus the concept of the weighting function as the vertical derivative of the transmission has application in the study of radiative flux in given spectral intervals and need not be restricted to vertical, monochromatic intensities. In the following we use hypothetical weighting functions in this way to elucidate, qualitatively, the different effects which increased CO_2 content has on the radiative flux in spectral regions of differing absorption strength.

The direct radiative influence of increased CO_2

Fig. 20(a) is an illustration of the type of weighting function profile associated with a strongly absorbing spectral interval near the centre of the 15 µm CO_2 band, together with two profiles for progressively less strong absorption further from the band centre. The broken profiles indicate the same weighting functions for an atmosphere with increased CO_2 concentration. Fig 20(b) shows equivalent weighting functions for an atmospheric path extending only to the tropopause, while Fig. 20(c) and (d) show corresponding weighting functions for downward radiation at the tropopause and at the surface. A typical atmospheric temperature profile is drawn to the right of each set. Note that the weighting functions illustrated are purely schematic and have not been drawn to scale.

Consider first the case of strong absorption. Here the paths from the surface to the tropopause and to the top of the atmosphere are opaque - i.e. the path transmissions are zero. An increase in CO_2 concentration therefore does not increase the area under the weighting function curve since

$$\int_{z_s}^{\infty} \frac{\partial \tau_{\bar{\nu}}^{z_s,\infty}}{\partial z}\, dz = \int_0^1 d\tau_{\bar{\nu}}^{z_s,\infty} = 1 \qquad \ldots (18)$$

Equation (16) indicates that if the atmosphere were isothermal there would be no change in the upward or downward radiative flux. However the increased concentration gives rise to an increase in the contribution from closer levels and a reduction in that from more distant levels. This may be visualized as a shift in the weighting function. For a non-isothermal atmosphere, therefore, the change in flux depends on the way in which the shift of the weighting function relative to the Planck function profile alters the contributions $B(T,z)K(z)\delta(z)$ from each level. Typically atmospheric temperature, and hence Planck function, decreases from the surface to the tropopause and increases with height throughout the overlying stratosphere. The upward flux in a strongly absorbing spectral interval at the top of the atmosphere must therefore increase since the upward shift increases the contribution from high levels in the strato-

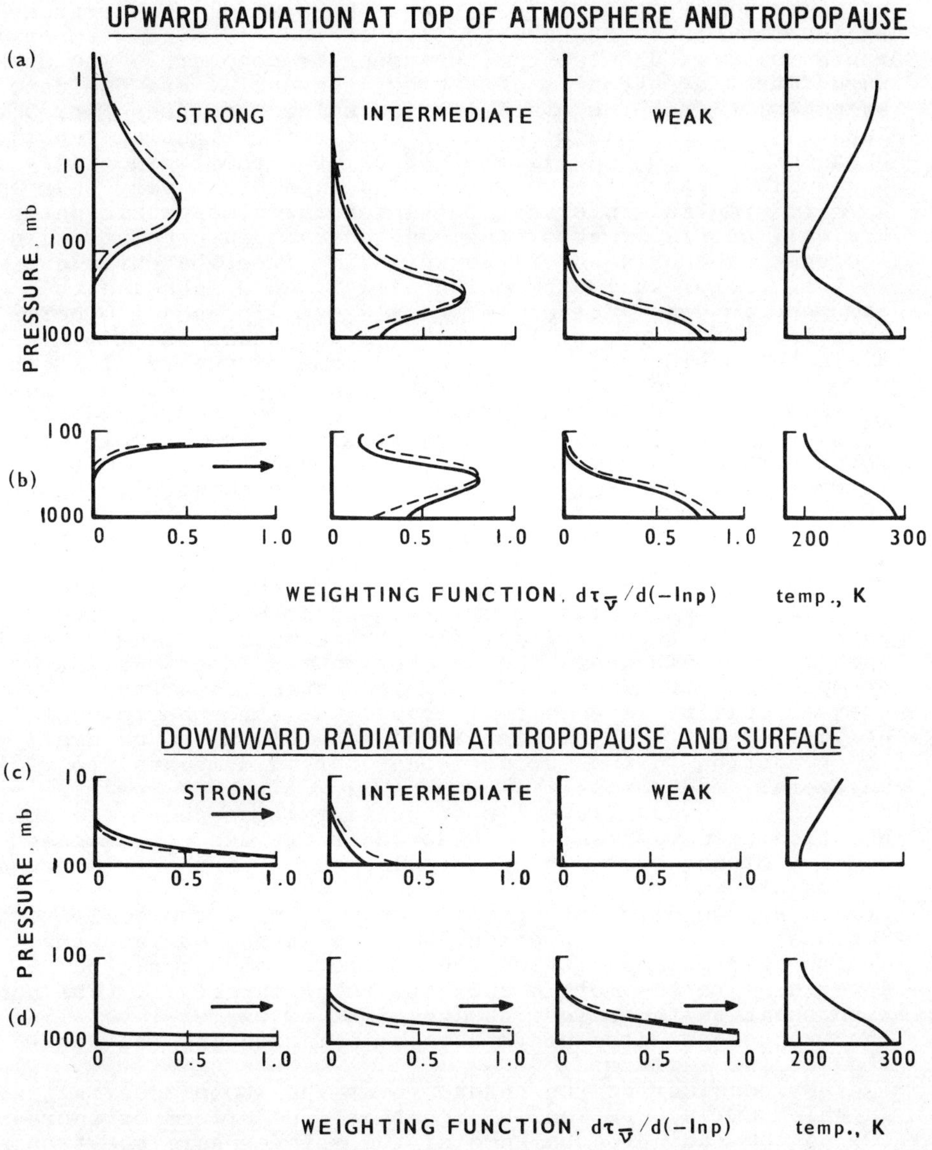

Figure 20. Hypothetical weighting functions for the upward radiative flux at (a) the top of the atmosphere and (b) the tropopause, and the downward flux at (c) the tropopause and (d) the surface, for three spectral intervals of strong, intermediate and weak absorption in the CO_2 spectrum. The broken profiles indicate the weighting functions for an increased CO_2 content. The natural logarithm of the pressure is used as the vertical coordinate. A typical vertical temperature profile is shown on the right.

sphere, where temperature is high, and reduces the contribution from the cooler upper troposphere and lower stratosphere. At the tropopause the upward flux is reduced since temperature falls with height in the upper troposphere. The downward fluxes at strongly absorbing frequencies are dominated by emission from the overlying few kilometres (see Fig. 20(c) and (d)) and the slight downward shift in weighting functions upon doubling CO_2 has little effect upon them, especially at the surface where the weighting function peaks very sharply.

In spectral intervals for which the atmospheric paths are only partially absorbing, an increase in concentration increases the area under the weighting function curve and, in the case of an isothermal profile, would increase the atmospheric contribution to the fluxes. For non-isothermal profiles the effect of the shift in weighting function must again be considered. Also, in the case of upward flux, the reduction of the surface contribution due to reduced transmission of the path (the first term on the right hand side of equation (16)) must be calculated. In the example of absorption of intermediate strength shown in Fig. 20 the weighting functions for the upward flux peak within the troposphere. Although there is an increase in emission due to increased opacity, this is more than offset by the upward shift of the weighting function to cooler levels. Coupled with the reduction in the surface component, the total reduction in upward flux is greater at these spectral frequencies, for which the path is partially absorbing, than in spectral regions where the absorption is either much stronger or much weaker. In the case of weak absorption, the increased opacity is seen to increase the atmospheric contribution from all levels (Fig. 20(a) and (b)) although again the reduction in the surface contribution dominates to yield an overall reduction.

The downward fluxes in spectral regions where the absorption is not strong increase generally due to increased opacity of the overlying path. Increases at the tropopause are mainly in spectral intervals for which the overlying stratospheric path is partially opaque, i.e. those close to regions of strong absorption, whereas at the surface the optimum partial opacity of the complete atmospheric path appertains to the more weakly absorbing intervals. The downward shift of the weighting functions to warmer levels in the troposphere also plays a role in increasing the downward flux at the surface.

By considering the changes over the whole spectral range in which CO_2 absorption is important, the effect of increased CO_2 on the radiation balance of the stratosphere the troposphere and the surface can be ascertained (remembering that at the moment consideration is being given to the initial radiative perturbations and not to subsequent feedback processes). At the top of the atmosphere, the increased emission of radiation in strong absorption intervals, which has originated mostly in the stratosphere, is more than offset by the decreased emission over the greater range of frequencies where the absorption is less strong. The Earth-atmosphere system therefore warms. From the earlier discussion it is clear that at the tropopause the upward flux decreases across the whole spectral range, and the downward flux increases in a significant portion of it. The combination of these changes leads to warming of the surface-troposphere system, and, since the total reduction of upward flux at the tropopause (i.e. reduced upward plus increased downward flux) is larger than that at the top of the atmosphere (taken here as the stratopause), the stratosphere suffers radiative divergence and must cool. These effects have been discussed quantitatively by numerous authors but in particular by Ramanathan et al. (1979) and Schneider (1975), who also consider the final thermal equilibrium states which evolve (see Chapter 6).

The division between tropospheric warming and surface warming is very dependent on the water vapour concentration in the troposphere. If there were no tropospheric absorbers with absorption spectra overlapping the CO_2 spectrum, the increase in downward flux at the surface would be larger than that at the tropopause, and the troposphere would cool. However the water vapour present renders the troposphere opaque to substantial portions of the CO_2 15 μm absorption band with the result that, at these frequencies, changes in the downward flux at the surface upon increasing CO_2 are initially small (prior to any water vapour feedback effects). The total increase in downward flux at the tropopause (i.e. increased downward plus reduced upward flux) is thus larger than the increase in downward flux at the surface, so both troposphere and surface warm. The model calculations of Ramanathan et al. (1979) indicate that where water vapour concentration is largest, i.e. in the tropics, the tropospheric warming tends to exceed the surface warming, while near the poles the reverse holds.

The direct radiative effects of increasing CO_2 discussed in this chapter are well established theoretically and to a considerable extent are capable of verification by laboratory measurements. Moreover, they are in agreement with observation in the sense that they are used successfully in the deduction of temperature profiles from atmospheric infrared emission measurements made from satellites. In contrast, the magnitude and relative importance of the various feedback mechanisms which may be brought into play following these CO_2-induced radiative perturbations are much less certain. Something of the way in which some atmospheric feedbacks operate can be learned from the study of the seasonal cycle of the present-day climate. Others are probably better analysed in the context of numerical simulations of the climatic response to higher carbon dioxide concentrations, although the complex, interactive roles of clouds and the oceans are still not sufficiently understood for their proper representation. Nevertheless, using numerical models of climate it is now possible to make an assessment of at least the average global, and to a lesser extent latitudinal, changes in climate that are likely to follow increases in carbon dioxide concentration. It is assumed that at some time in the forseeable future (see page 95) the carbon dioxide concentration will attain a value twice that of the present day, and thus elucidating the effects of a doubling has been adopted as a standard for CO_2-climate studies. The uncertainties behind the estimates of the magnitude and timing of future carbon dioxide levels have been discussed earlier and are of no direct concern here. In the following three chapters we discuss the use of climate models in this context and address some of the major difficulties involved.

6 One-dimensional Climate Models

Numerical models of various degrees of complexity have been used to investigate the effects of increased carbon dioxide concentrations. Here we shall discuss the results of the relatively simple one-dimensional models. Chapters 7 and 8 deal with three-dimensional general circulation, or global climate, models (GCMs).

Horizontally-averaged vertical column models

One-dimensional vertical column models are used to study the globally averaged, annual mean temperature profile and can yield a first estimate of the likely change in the globally averaged temperature profile due to a doubling of carbon dioxide. Certain important feedback processes cannot be modelled, including those between radiation and dynamics and those dependent on latitude, such as the ice albedo-temperature feedback. The temperature profile is determined by the vertical distribution of solar and infrared radiative fluxes which are controlled by the specified vertical distributions of optically active atmospheric gases, aerosols and clouds and by the reflective and scattering properties of these components and the Earth's surface. Usually the temperature lapse rate in the troposphere is adjusted to allow for the effects of convection, since radiative processes alone yield markedly unstable temperature profiles in the lower atmosphere. The limitation to one dimension permits fairly detailed modelling of the physical processes involved, and the parametrization schemes of these 'radiative-convective' models are often adopted, sometimes in simplified form, in GCMs. An excellent review of the use of radiative-convective models for climate modelling is given by Ramanathan and Coakley (1978).

A number of CO_2-climate sensitivity studies have been performed using radiative-convective models, e.g. Manabe and Wetherald (1967), Schneider (1975), Augustsson and Ramanathan (1977). Had the weak carbon dioxide absorption bands been included in the radiation schemes of all these models then all would have computed an increase in surface temperature of about 1.9 K following a doubling of carbon dioxide. Implicit in these calculations are the direct radiative effects of carbon dioxide and the water vapour-temperature feedback. The latter arises as a result of increased evaporation from the warmed surface and the ability of the warmer lower atmosphere to hold more water vapour. The additional water vapour causes further 'greenhouse' warming which contributes, in the global mean, about equally with carbon dioxide to the overall increase of about 1.9 K.

Ramanathan et al. (1979) used a radiative-convective model to evaluate the latitudinal and seasonal variation in the direct radiative flux changes arising from doubling carbon dioxide. Their results indicate that the excess heating is larger at the equator than at the poles, and in high latitudes is larger in summer than in winter. These variations arise from the temperature dependence of the carbon dioxide radiative processes. As discussed on page 57, overlap of the water vapour and carbon dioxide infrared spectra leads to most of the heating occurring within the troposphere in low latitudes where water vapour is abundant, with much smaller direct radiative heating of the surface. Here surface heating results from enhanced downward infrared flux from the warmed troposphere and the water vapour-temperature feedback. In drier, high latitudes the surface warming due to doubling carbon dioxide is much greater than the warming of the overlying air.

Following on from this work, Ramanathan (1981) has analysed quantitatively the individual processes involved in the surface-troposphere energy exchange consequent upon doubling CO_2 in a one-dimensional radiative-convective atmosphere-ocean model. Fig. 21 illustrates the direct and feedback warming mechanisms. Since the relative contributions of the various processes to surface warming are to some extent model-dependent, we shall not discuss the figure quantitatively except to note that process 3, constituting the water vapour feedback and latent heat loops, is the major surface warming agent.

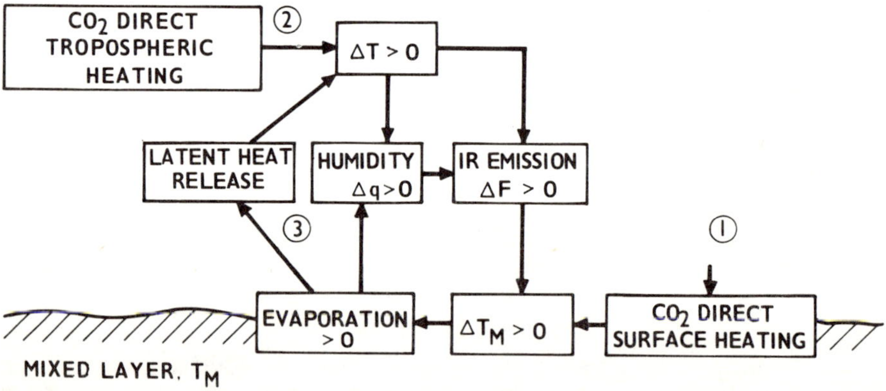

Figure 21. Direct and feedback processes consequential upon increased carbon dioxide. Process 1 is direct radiative heating of the surface, process 2 involves direct tropospheric heating and consequent enhancement of surface heating, while process 3 constitutes the water vapour feedback and latent heat release. (Adapted from Ramanathan, 1981)

It is important to obtain some idea of how sensitive the change in modelled surface temperature is to assumptions relating to the humidity profile, the lapse rate and the characteristics of clouds. In the models discussed so far, cloud-top altitudes were fixed. Augustsson and Ramanathan (1977) performed the same calculation except that a fixed cloud-top temperature and a temperature-dependent, rather than a fixed, relative humidity were employed in turn. For constant relative humidity and fixed cloud-top temperature a surface temperature increase of 3.2 K occurred. Reverting to fixed cloud-top altitude, values of 2.6 K and 3.0 K were computed for two different formulations of the dependence of relative humidity on temperature. These results compare with 2.0 K for fixed relative humidity. If cloud-top temperature is fixed, the effective radiating temperature of the cloud is lower than when the cloud-top altitude is fixed, since in the latter case overall tropospheric warming produces warming of the cloud-tops following the doubling of carbon dioxide. Fixed cloud-top altitude therefore results in greater infrared cooling and a smaller rise in surface temperature than in the fixed cloud-top temperature case. There is as yet no theoretical basis for preferring either cloud-top assumption or any particular formulation for the temperature dependence of relative humidity.

As noted above, since the radiatively-induced temperature changes may give rise to unstable lapse rates in the troposphere, a radiative-convective model is designed to simulate in some way the tendency for convection to restore a stable vertical temperature structure. The variety of 'convective adjustment' schemes which have been used further widens the span of uncertainty in the overall surface warming due to increased CO_2. Rowntree and Walker (1978) demonstrated that the use of a penetrative convection scheme, based on buoyancy principles, produces a tropospheric lapse rate close to the moist adiabatic lapse rate which becomes less steep with increasing temperature. Thus, a given surface warming is associated with a larger warming in the upper troposphere. As a result the infrared cooling from the upper troposphere is larger than obtained with a model employing a fixed tropospheric lapse rate. This additional cooling is sufficient to outweigh the stronger water vapour feedback in this scheme and a surface warming significantly less than for the fixed lapse rate model is obtained, especially for temperatures characteristic of warmer latitudes. Not surprisingly the results are quite sensitive to assumptions made about the dependence of relative humidity on temperature.

These results clearly indicate the large sensitivity of a CO_2-induced surface warming to certain physical processes occurring in the troposphere, about which knowledge is incomplete. Since the same schemes are often adopted in GCMs, the radiative-convective model studies give a useful indication of the differences that should be expected between GCMs using different parametrizations.

Although the values for the surface warming given by some radiative-convective models are in quite good agreement with the globally averaged values deduced from GCMs, their results cannot be regarded as being predictive since all regional and latitudinal variations and the associated feedback processes are excluded.

Latitudinally varying energy balance models

A first approximation to the spatial distribution of a CO_2-induced warming may be ascertained from energy balance models (EBMs). These models calculate the latitudinal variation of surface temperature in terms of a balance between incoming and outgoing radiation and the latitudinal fluxes of sensible and latent heat. They may be time-dependent allowing seasonal variations to be included or be time-invariant for annual mean conditions. The radiative processes must represent averages for the vertical column. The outgoing infrared flux is often parametrized in terms of the surface temperature and cloud cover based on empirical data. In some models the atmospheric temperature profile is specified as a function of surface temperature. The lapse rate may then vary with latitude and season so that radiative processes may be better treated. The dynamic heat fluxes must also be expressed in terms of a surface temperature parameter; the local meridional gradient of temperature or the departure from the hemispheric mean temperature are often employed. The albedo is best formulated as a function of surface temperature (so that the different albedos of ice and water may be represented), cloud amount and zenith angle. Some early EBM studies interpreted the empirically observed dependence of albedo on zenith angle as a temperature dependence and consequently over estimated the importance of the ice albedo-temperature feedback - the sequence whereby surface warming reduces ice cover, lowers the albedo, increases solar absorption and hence enhances further surface warming.

Ramanathan et. al. (1979) used a radiative-convective model to determine the change in net radiative flux at the tropopause due to doubling carbon dioxide as a function of latitude and season. The EBM of Lian (1978) was then employed to evaluate the latitudinally and seasonally varying response of the surface temperature to these changes in radiative input. The results, for the northern hemisphere, gave a change in the hemispheric annual mean surface temperature of 3.3 K. At low latitudes little seasonal variation from a mean near 3.1 K occurred. In the Arctic, maximum increase in temperature was found in Spring and Summer. At 85°N the change was 12 K in summer and 4 K in winter. This high latitude variation was regarded as being due to ice albedo-temperature feedback rather than the seasonal variation of the increased infrared radiation. The authors pointed out, however, that Sellers' (1974) EBM produced a maximum high latitude warming in mid-winter, presumably arising from other feedback mechanisms. The influence on the high latitude seasonal variation of CO_2-induced warming may also be strongly influenced by the behaviour of sea ice, as discussed later, a feature not included in the above model.

It is worth noting that considerable doubt has recently been expressed concerning the validity of EBMs. The parametrizations of the energy budget components are generally derived from observations of the present climate and are designed so that the model reproduces the observed temperatures and energy fluxes. Coakley and Wielicki (1979) used the results of a GCM as data for formulating the parametrizations of an EBM and compared the response of both the GCM and EBM to a change in climatic forcing (changes in the solar constant). Changes in surface temperature in the EBM were found to be much larger than in the GCM, indicating that the feedbacks which governed the response of the GCM were not present in the EBM. They concluded that if an EBM cannot model the response of a GCM to a change in climatic forcing, it is unlikely to be capable of modelling real climatic change.

7 Global Climate Models

Introduction to GCMs

For more detailed analyses of CO_2-induced warming than given by the one-dimensional vertical and horizontal models, recourse to the three-dimensional GCMs has been made.* In these models the current values of the meteorological variables which define the state of the atmosphere are either held at the intersection points of a three-dimensional grid (finite difference models) or represented in terms of spherical harmonics on each of a set of levels in the atmosphere (spectral models). The horizontal resolution in the latter type of model is determined by the number of harmonics retained, whereas that in the finite difference model depends on the grid spacing.

The models are based on the set of mathematical equations which govern the physical behaviour of the atmosphere (the three equations of motion, the thermodynamic equation, the equations of conservation of mass and water and the equation of state). Given an initial state of the model atmosphere and appropriate boundary conditions, the model is integrated forward in time steps of perhaps ten minutes of model time to yield the atmospheric state at future times.

The physical processes which affect the meteorological variables in reality cover a range of temporal and spatial scales spanning at least ten orders of magnitude. The most slowly varying components of the system may be regarded as constant and so pose no problems. However to model the important small scale processes explicitly would involve so large a computational expense, due to the fine resolution required, that a sufficiently long integration to be useful would not be possible with present-day computer power. The

* Recent suggestions (Newell and Dopplick, 1979; Idso, 1980) that the estimates of CO_2-induced warming using GCMs are an order of magnitude too large are unfounded. The simple surface energy balance approach adopted by these authors, and on which their own assessments of CO_2-induced warming were based, failed to take account of CO_2-induced tropospheric warming (see Chapter 5) and ignored some of the basic radiative and dynamical feedback processes involved, such as the important water vapour feedback. More detailed discussions of the shortcomings of these papers have been given by Watts (1980, 1982), Ramanathan (1981), Crane (1981) and National Academy of Sciences (1982).

philosophy adopted is to choose a horizontal resolution that can just resolve the smallest meteorological processes of primary interest. A higher spatial resolution not only involves more computation directly but also requires that a shorter time step be used to ensure computational stability. The stability limitation on the time step may be removed by using implicit integration schemes, although much of the advantage may be lost since these are computationally more time-consuming.

For global climate models in which integrations of several years are necessary the horizontal grid must resolve the synoptic scale motions (anticyclones and depressions) and usually has a spacing of a few hundred kilometres. This means that all processes of smaller scales which are known to interact in important ways with the resolvable processes and which must be represented in the governing equations, such as cumulus convection, cloud formation and dissipation, precipitation, exchanges of heat, moisture and momentum at the surface, etc., must be parametrized - i.e. the average effect of these 'subgrid-scale' processes within an elemental volume of the grid must be expressed in terms of the variables held at the grid scale. The development of satisfactory parametrizations is often the most difficult aspect in the development of climate models.

The main use of GCMs in CO_2 research has been in carrying out sensitivity experiments. In such experiments two integrations of the model are performed. First the model is run using present-day climatic boundary conditions from a prescribed initial state (which may, for example, be merely a stationary, isothermal atmosphere) until a statistical equilibrium climate is achieved. At this point the initial secular trends in the evolution of the model climate will have long since disappeared and all the main features of the real climatic system will be recognizable. The term 'statistical equilibrium' implies that the model will not be in a steady state in the mathematical sense but will exhibit day to day variability just as the observed climate does. A quantitative description of this equilibrium climate is obtained by averaging over a suitable period towards the end of this first integration, which is often termed the control run. A second integration is carried out in which a specified change in a boundary condition, such as a doubled CO_2 concentration, is imposed. This is run until statistical equilibrium is reached. The differences between the statistical equilibrium climates obtained in the control and perturbed runs are regarded as indicative of the changes which would occur if the prescribed change in boundary condition occurred in the real climate system. The length of the integration required to attain statistical equilibrium depends on the complexity of the model. Typically a seasonally-varying model using climatologically observed sea surface temperature may require about four model years, whereas one in which the thermal characteristics of the surface mixed layer of the ocean are included needs some 10-15 years.

Before discussing GCM carbon dioxide doubling experiments in detail (Chapter 8) it is instructive to examine some of the major deficiencies in present-day GCMs. Already mentioned is the fact that insufficient computer power prevents modellers including in their models all the features which they could include. For example, highly idealized geography and no seasonal variation may be necessary if the advantages of a finer resolution are chosen. Often these simplifications are forgotten when model results are discussed. Another limitation of current modelling studies is that none of the models is yet capable of addressing properly the time-dependent nature of gradually increasing carbon dioxide concentrations, simulation of which is essential for gaining insight into the evolution of the future climate (see page 83). Again this is at least partially due to inadequate power of currently available computers. Probably the two most fundamental outstanding problems in the present generation of models, however, are how to model clouds and how to couple the ocean to the atmosphere in a way which allows proper thermal interaction between them. These aspects of climate modelling are sufficiently important to merit fuller discussion and each is addressed in turn below.

The representation of clouds in GCMs

Clouds interact with several features of the climate system: they couple the hydrological processes of the atmosphere and the ground through precipitation, through their formation and dissipation they redistribute latent heat, and they play an important role in reflecting, absorbing and emitting radiation.

Many climate models have used climatologically observed cloud distributions so that their radiative influence may be incorporated. In this case the hydrological and radiative processes are not linked. Condensation is assumed to produce rainfall directly without passing through a cloud phase. Such a scheme is obviously not ideal since changes in the temperature and humidity structure of the atmosphere, consequent on a carbon dioxide increase, are likely to lead to changes in cloud amount and distribution which further change the radiation budget and influence atmospheric motion through latent heat processes. In order to model these feedback mechanisms, interactive cloud schemes must be adopted.

The effect of clouds on the Earth's radiation budget is complex. The presence of cloud leads to greater reflection of solar radiation so less energy is available to heat the earth-atmosphere system. At the same time clouds absorb infrared radiation emitted from the surface and air beneath. The net effect of these competing processes depends largely on the cloud-top altitude. Simple reasoning suggests that low clouds, with cloud-top temperature close to that of the surface, emit radiation upwards almost as effectively as the surface, so do not significantly reduce the infrared loss from the lower atmosphere. High clouds, whose tops have low temperatures, emit infrared radiation much less intensely and consequently exert a stronger 'greenhouse' warming on the surface and atmosphere beneath. The albedo of high cloud is generally less than that of low cloud. It is thought that for high clouds the warming influence probably exceeds the relatively weak cooling due to solar reflection and they thus produce an overall warming. Surface cooling due to reflection of solar radiation by low clouds, however, far exceeds their slight infrared warming effect.

A number of observational studies involving satellite measurements of albedo and outgoing terrestrial infrared radiation have attempted to assess on a regional basis the net effect of cloud cover on the Earth's radiation budget. Recent results (Ohring and Clapp, 1980; Hartmann and Short, 1980) indicate that over the globe as a whole the reduction in solar radiation exceeds the reduction in infrared output; thus in the present climate clouds have an overall cooling effect on climate. The reduction in infrared output appears to dominate only over high latitudes in winter, where solar input is small, and over glaciated land masses in summer. From these observations it is clearly important that any changes in cloud amount and distribution accompanying tropospheric warming should be properly modelled in climatic change simulations. Unfortunately, there are insufficient observational data for validating interactive cloud schemes, which, though they may be tuned to give a good description of present-day cloud distribution, may not necessarily respond correctly to the warming associated with increased carbon dioxide.

The modelling of clouds is difficult since nearly all clouds are sub-grid scale, either horizontally, in the case of convective cloud, or vertically, for all but the thickest stratiform cloud. Even thick stratiform cloud may be banded and only marginally resolvable horizontally. Thus the dynamical and physical processes involved in their formation and dissipation are not resolved in the models and the cloud characteristics of any grid box must be parametrized in terms of the variables of the resolvable scale held at the grid points.

Several papers have been published which describe numerical experiments in which GCMs incorporating interactive cloud schemes have been used to assess the sensitivity of cloud amount and distribution to changes in surface temperature. Schneider et al. (1978) employed a model in which cloud was predicted at two distinct levels. Fractional cloud cover was computed as a function of the grid-scale relative humidity, subject to a vertical velocity criterion and maximum and minimum permissible cloud cover fractions. Prescribed changes of surface temperature were imposed either globally or in zonal strips on a fixed January climate. The cloud response was highly dependent on the latitudinal distribution of the surface temperature change and strongly related to local vertical motions. A deficiency of real data precludes verification of the feedback processes which occurred in the model.

Manabe and Wetherald (1980) carried out a CO_2-doubling experiment in which 80% cloud cover was imposed whenever condensation occurred. The results of this work will be discussed further below; only the response of the cloud to the increase in carbon dioxide is noted here. Cloud cover decreased in the upper and middle troposphere at most latitudes, but increased near the surface and in the model lower stratosphere in high latitudes. Both cloud cover and effective cloud top height were thus reduced equatorward of 50° latitude giving roughly compensatory changes in the solar and infrared radiative heating of the surface. At high latitudes, although the effective cloud top height remained at about the same level, the albedo and infrared effects were again roughly compensatory owing to the weak insolation in high latitudes. The authors stressed that, in view of the very crude cloud parametrization scheme used and uncertainties in optical properties of clouds, it was premature to conclude that the changes in cloud cover associated with carbon dioxide warming would neither increase nor reduce the CO_2-induced climatic change.

These results can be compared with similar experiments carried out by Hansen at the NASA Goddard Institute for Space Studies (GISS). Although the latter work is unpublished some details of the results were given in a report of the US National Academy of Sciences (National Academy of Sciences, 1979) to which further reference will be made in the next chapter dealing with CO_2-climate sensitivity experiments. In the GISS model, the method of parametrizing cumulus convection was different from that used by Manabe and Wetherald in that it permitted convection to extend through unsaturated, stable layers and did not allow for entrainment of cloud-free air. Convective cloud therefore extended to greater heights in the GISS model. As a result, doubling carbon dioxide led to an increase in high cloud at low latitudes, in contrast to the reduction obtained by Manabe and Wetherald. At high latitudes the Manabe and Wetherald model, which had no seasons, gave an increase in both high and low cloud, whereas the GISS model, which did have seasons, predicted an increase in high cloud throughout the year but an increase in low cloud only in spring. It is not known if the methods of predicting stratiform cloud differed between the models, but it would appear that the inclusion of seasonal variation was responsible for the different low-cloud predictions for high latitudes. The predominance of increasing high cloud in the GISS model must have given rise to an overall positive cloud-temperature feedback, with additional surface warming.

On the basis of a one-dimensional radiative-convective model study, Hunt (1981) questioned whether sufficient accuracy in model-generated cloud properties would, or could, ever be achieved. His results showed that a modest increase in low cloud cover from 0.30 to 0.33 could completely compensate for the surface and tropospheric warming which would be induced by a doubling of CO_2. Similar compensation was realised by increasing low cloud albedo from 0.69 to 0.74.

Paltridge (1980) has emphasized that cloud amount is but one cloud characteristic that could feed back on climate, and others, such as cloud water content, albedo and emissivity, have not been considered as variables in GCM cloud parametrization schemes. Using a one-dimensional radiative-convective model he demonstrated that an increase in cloud water content, which might be expected in a moister atmosphere, could produce a feedback as effective in reducing a CO_2-induced warming as an increase in low cloud cover. Stephens and Webster (1981) further examined the influence of changes in cloud water content on the radiative properties of clouds and also drew attention to the dependence upon surface albedo of the sign and magnitude of surface temperature changes due to transitions between clear and cloudy skies. Whereas increased cloud cover may produce cooling at a poorly reflecting underlying surface, warming may occur for the same change in cloud cover at a surface with high albedo.

It is abundantly clear that the problem of assessing both the response of cloud characteristics to an increase in CO_2 concentration and the effect of that response upon surface temperature is far from being resolved, and will remain an area of major uncertainty for a considerable time.

The representation of the oceans in GCMs

The oceans have three major influences on climate. They are the source of moisture for the hydrological cycle, they act as a thermal regulator, reducing the amplitude of the seasonal atmospheric temperature variation by virtue of their large thermal capacity, and their currents transport heat horizontally. Weaknesses in our present understanding of the dynamics of the oceans have limited the development of dynamical ocean models. Difficulties are posed by the occurrence of mesoscale phenomena which contribute significantly to oceanic transport but which are an order of magnitude smaller in size than the synoptic circulation features in the atmosphere. The dynamical nature of these eddies is not well understood, and, even if it were, the horizontal resolution required to model them would be prohibitively high. Knowledge of the processes involved in vertical mixing is equally poor (see Chapter 2). Thus, relative to the description of the atmosphere, the ocean in most models of the climate system is represented only very crudely.

All GCMs used in climate studies must include an ocean surface that can supply moisture to the atmosphere, and this is not difficult. Rather more of a problem is the modelling of the ocean's thermal capacity. This is in part determined by the type of simulation being performed. Two simple, but extreme, methods are either to use climatologically observed sea surface temperatures (e.g. Gates et al., 1981; Mitchell, 1983) or to assume the oceans are merely a 'swamp' or wet surface (e.g. Manabe and Wetherald, 1975). In the first case the atmosphere is not coupled to the ocean. The ocean effectively has an infinite thermal capacity and cannot respond to atmospheric temperature changes. The swamp model, conversely, has zero heat capacity. If a CO_2-doubling experiment is to be carried out in which there is no seasonal variation and in which both the control and doubled-CO_2 simulations are run to equilibrium, the exclusion of the ocean's thermal capacity does not matter since it would only serve to delay the attainment of the equilibrium state. A swamp model would thus be preferred over the employment of climatological sea surface temperatures unable to respond to warming of the atmosphere (see later comparison on page 75).

If the seasonal variation of CO_2-induced warming is being investigated it is essential to use a model which takes account of the thermal capacity of that part of the ocean - the surface mixed layer - which responds to the seasonal temperature cycle. In its simplest form the mixed-layer ocean model is merely a vertically isothermal slab of water of appropriate uniform depth to simulate the seasonal thermal behaviour of the real ocean. The thermal capacity of the deep ocean is not important in influencing the seasonal variation of atmospheric temperature.

In none of the schemes for incorporating the effects of the oceans so far discussed is the horizontal heat transport by ocean currents modelled explicitly. This is a serious deficiency since the oceans are believed to achieve between one third and one half of the total poleward heat transport in the present climate (Vonder Haar and Oort, 1973). While climate models employing observed sea surface temperatures cannot simulate reliably a CO_2-induced climatic change, they should be better at modelling the present-day climate than those with swamp or simple mixed layer oceans, since the effect of oceanic heat transport is inherent in the seasonally varying observed sea surface temperatures. In GCMs coupled with swamp or mixed-layer models, the oceanic poleward heat flux must instead be achieved by a more active atmospheric circulation. Thus, while the global heat balance is maintained, the regional climatic influences of ocean currents, such as the Gulf Stream, are absent. A doubling of carbon dioxide is likely to influence both the oceanic and atmospheric heat transports, and a model which allows only atmospheric transport can not be expected to simulate the detailed climatic consequences of these changes. Thus, both the regional and latitudinal distribution of CO_2-induced warming may be in error as a result of neglecting ocean dynamics.

Recently progress has been made in the coupling of three-dimensional oceanic GCMs with atmospheric GCMs (e.g. Manabe et al., 1979; Washington et al., 1980). These models do not resolve the finer details of the oceanic currents but appear capable of reproducing the larger-scale features of the oceanic circulation and their seasonal variation. A tendency for modelled oceanic heat transports to be less than those estimated from observation has been noted, probably arising from too coarse a resolution. Coupled GCMs are still at an early stage of development and have yet to be used in climate sensitivity experiments, although one such model has been applied to the problem of estimating the thermal lag which the oceans would exhibit in the event of climatic warming (see page 83).

One of the major difficulties involved in joint atmosphere-ocean modelling is devising suitable methods for coupling the two component models. The problem arises largely as a result of the very different thermal inertia of the atmosphere and oceans. To reach thermal equilibrium a complete ocean model starting from rest must be integrated for periods of order 10^4 years, whereas integrations of less than 10 years are adequate for the atmospheric model. If the ocean is represented by a mixed-layer model only, the disparity of thermal relaxation times is much less but still poses the same problem. The computer time required to integrate the atmospheric model over a given period is typically a few hundred times larger than the equivalent integration time for the ocean owing to the greater complexity of atmospheric processes. Consequently, using the short time steps demanded for computational stability of the atmosphere, it is prohibitively expensive to perform synchronous integration of the joint evolution of the atmospheric and oceanic circulations which are sufficiently long for the ocean to obtain thermal equilibrium. Schemes in which the two components are integrated asynchronously have therefore been developed.

Manabe et al. (1979) coupled a 1200-year ocean integration with an atmospheric integration of 4.2 years, each component making available to the other continuously updated information on the variables relevant to the energy exchange processes between them. In the GCM at the National Center for Atmospheric Research (NCAR), Washington et al. (1980) used sample 10-30 day integrations of January, April, July and October conditions in the atmospheric model to derive annual and semiannual harmonics of the atmospheric forcing of the ocean with which to drive the ocean model for a five-year period, after which the new (seasonal) ocean surface characteristics were employed in the next sample atmospheric integrations. A not dissimilar interactive approach was adopted in Manabe and Stouffer's (1980) mixed-layer model. Their technique for reducing atmospheric integration time was to employ, initially, an accelerated atmospheric seasonal cycle. Towards the end of the joint integration the seasonal cycle was extended to a year so that a realistic atmospheric climate would evolve.

The above discussion has been concerned primarily with modelling the ocean in sensitivity experiments which are run to equilibrium. The intermediate 'climates' along the road to equilibrium are of no interest, and, especially if an asynchronous coupling scheme is used, will not be representative of any actual climatic transition. It will be recalled that to simulate the seasonal variation of climate a model must contain an oceanic mixed layer so that the seasonal lag introduced by the thermal storage of the upper ocean is simulated. Similarly, if it is wished to model the evolutionary response of the climate system to a step-perturbation or a time-dependent forcing, the thermal lag introduced by the whole ocean must be simulated. This requires long-term synchronous integration of a coupled atmosphere-ocean model in which the three-dimensional oceanic circulation is included. Preliminary studies which have been made towards determining the transient response of the climate system to increasing CO_2 concentrations are discussed on page 83.

8 Application of Global Climate Models to the Carbon Dioxide Problem

In the previous chapter we have highlighted some of the difficulties involved in modelling the physical processes of the climate system. Numerous other problem areas could be identified. Indeed the descriptions of individual physical processes in GCMs are often so crude that it is surprising that the models achieve any success at all! It is not unusual, therefore, to find that while several models may yield similar zonal averages of meteorological variables in a doubled-CO_2 climate, features largely controlled by the broad scale distribution of radiation and land-sea differences, their regional climate simulations may differ considerably. Thus while some analyses of regional changes in model climate due to doubling carbon dioxide have been carried out, their authors have been the first to stress that the results are not regional predictions but merely indicate that one should expect any climatic change to be far from uniformly distributed. The improvement of the capability of models to predict regional climate and its variability, which will come mainly from progress in modelling clouds and the oceans, is widely recognised as the major objective of general circulation modelling. An important aspect of this work will be to verify that the models are able to simulate not only mean climatic conditions, but also inter-annual climatic variability, the realisation of which will require extended model integrations of many years.

In the following are summarized the results of the main CO_2-doubling (or quadrupling) sensitivity experiments carried out using GCMs. These are relatively few in view of the large expense involved in running these models. An Ad Hoc Study Group was assembled by the Climate Research Board of the US National Research Council in July 1979 to make a scientific assessment of the carbon dioxide-climate problem (National Academy of Sciences, 1979). They compared in detail the five models which they considered to be the most realistic so far developed. Three of the five were models developed by Manabe and colleagues at the NOAA Geophysical Fluid Dynamics Laboratory (GFDL), the other two by Hansen and colleagues at the Goddard Institute for Space Studies (GISS). Only the results of the GFDL experiments have so far been published in the mainstream literature, so the details of the GISS model computations are extracted directly from the Study Group's report. Since that report was issued two further CO_2-climate sensitivity experiments at GFDL have been published as have the results of two other modelling groups at the British Meteorological Office and Oregon State University.

In the following discussion we deal first with simulated CO_2-induced changes in temperature and then the less extensively researched problem of changes in the global water budget.

CO_2-induced temperature changes

The first GCM CO_2-climate sensitivity study was that of Manabe and Wetherald (1975) at GFDL. The model's geography was idealized, consisting of repeating 60° longitudinal sectors of land and ocean (a sector model). Only one hemisphere, truncated at 81.7° latitude, was considered. A swamp ocean was used, sea surface temperatures being determined from the requirement of instantaneous thermal equilibrium with the overlying atmosphere. Cloudiness was fixed and there was no seasonal cycle. In a successor to this model (Manabe and Wetherald, 1980) the computational domain was extended to the pole, a finer horizontal resolution was employed and a simple, interactive cloud scheme (as discussed in Chapter 7) was introduced. In other respects the models were very similar or identical. Since the cloud-radiation feedback was found to be small these models produced very similar results. The hemispheric mean temperature increase for a doubled-CO_2 atmosphere was about 3 K with the zonal mean change (i.e. the average change around a latitude circle), shown in Fig. 22, ranging from less than 2 K at the equator

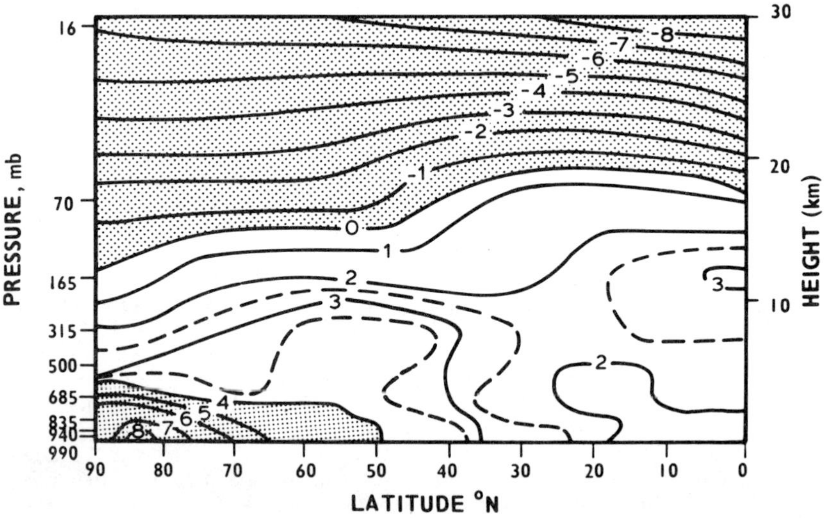

Figure 22. Latitude-height distribution of the change in zonal-mean temperature (K) in response to a doubling of CO_2 content as given by Manabe and Wetherald's (1980) annual mean sector model. (Reproduced from Manabe and Wetherald, 1980)

to about 8 K in high latitudes. The greater increase at high latitudes was due to enhancement of the warming by the retreat of highly reflective snow cover - the ice-albedo feedback, the confinement of additional heating to the lowest layers on account of the stable atmospheric stratification, and increased poleward transport of latent heat due to the overall warming and moistening of the atmosphere. Comparing the results of the doubled-CO_2 experiment with a similar quadrupled-CO_2 integration, it was shown that the area-mean increase in surface temperature was roughly proportional to the change in the logarithm of the carbon dioxide concentration, being close to 6 K for quadrupled concentration of carbon dioxide. This is a consequence of the warming effect being due principally to enhanced absorption in partially absorbing regions of the CO_2 spectrum (see Chapter 5 and Augustsson and Ramanathan, 1977).

It is interesting to note here the results of two studies carried out by the modelling groups at the British Meteorological Office (BMO) and Oregon State University (OSU) (Mitchell, 1983; Gates et al., 1981). These examined the effect of doubling carbon dioxide in models which effectively contained oceans with infinite heat capacity as opposed to the swamp models of Manabe and Wetherald described above. Both models employed monthly climatological sea surface temperatures, which were kept fixed when CO_2 was doubled, and incorporated realistic geography and a seasonal cycle. A doubling of carbon dioxide produced a global, annual average increase in surface temperature of only 0.2 K. Apart from preventing the air overlying the oceans from warming, these simulations did not allow the enhancement of the surface temperature increase due both to increased moisture content (the water vapour feedback) and the sea ice-albedo feedback - both discussed in Chapter 6. The results demonstrate the important influence of the oceans on the atmospheric response to doubling carbon dioxide.

As noted earlier, experiments with fully coupled atmospheric and oceanic GCMs have yet to be made. An advance towards this goal was achieved by Manabe and Stouffer (1979, 1980) using an atmospheric GCM underlain by a static, vertically-isothermal mixed-layer ocean. The model employed fixed cloudiness (annual mean climatological values) but a realistic geography and seasonally varying insolation. The ocean depth, 68 m, was chosen so that the heat storage associated with the annual cycle of observed sea surface temperature was correctly modelled. Ocean temperature changes were based on the surface heat budget and the varying thickness of sea ice was modelled. A control run was compared with a quadrupled-CO_2 integration. Fig. 23 shows the annual mean latitude-height distribution of the difference in zonal mean temperature between the two runs, while Fig. 24 shows the increase in zonal mean temperature at the surface as a function of season. The annually averaged global mean rise of about 4 K suggests an increase of 2 K for doubled CO_2 concentration, which is 1 K less than obtained by Manabe and Wetherald's model described above. The causes of the smaller response were: (i) less warming in southern than in northern high latitudes due to a much smaller ice-albedo feedback (see below) resulting in a smaller global mean

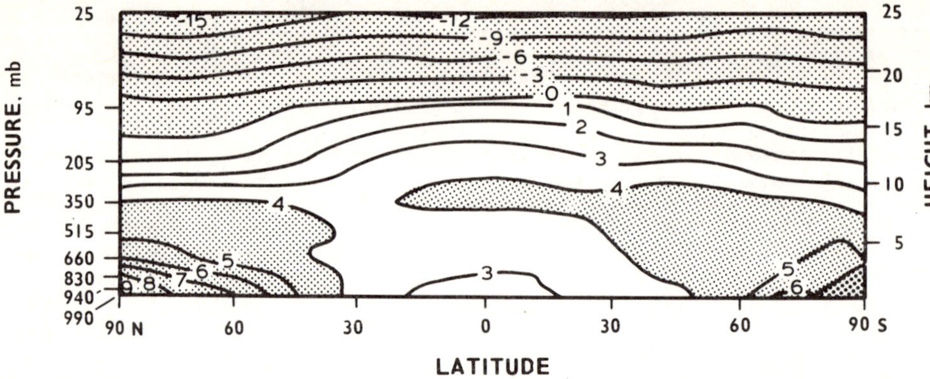

Figure 23. Latitude-height distribution of the zonal mean difference in annual mean temperature (K) between present and quadrupled CO_2 experiments with Manabe and Stouffer's (1979, 1980) global, seasonal mixed-layer model. (Reproduced from Manabe and Stouffer, 1980)

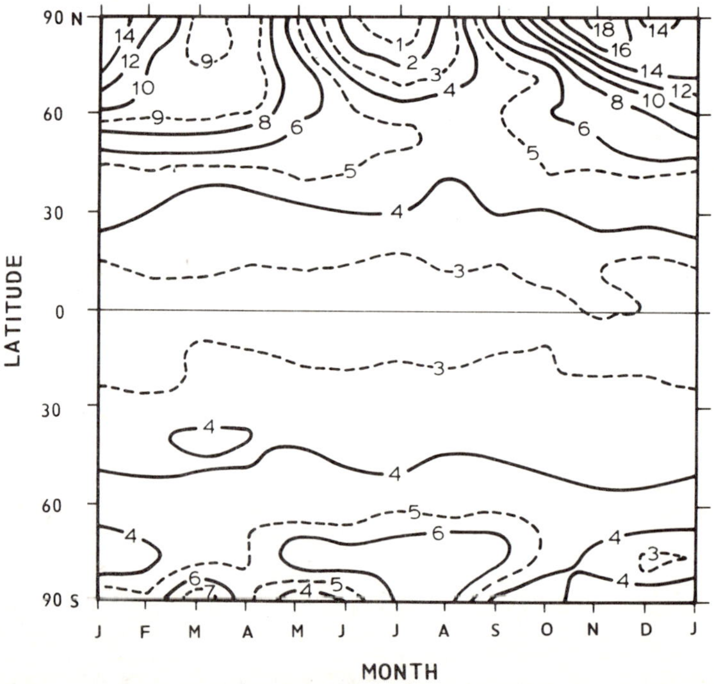

Figure 24. Latitude-time distribution of zonal mean difference in surface air (70 m altitude) temperature (K) between present and quadrupled CO_2 experiments using the same model as in Fig. 23. (Reproduced from Manabe and Stouffer, 1980)

temperature rise than in the single (northern) hemisphere model of Manabe and Wetherald, and (ii) less warming in high (mostly northern) latitudes due to the influence of the seasonal variation.

To investigate the less marked warming in high latitudes in the seasonal model, Wetherald and Manabe (1981) compared two atmosphere-mixed layer sector models, one with seasonally varying insolation and the other with annual mean insolation. The seasonal model simulated snow-melt in spring whereas the annual mean model retained a fixed annual mean snow cover. The CO_2-induced warming moved the snowline polewards in the annual mean model and a snow-albedo feedback therefore operated throughout the year. In the seasonal model, however, snow-albedo feedback was confined to winter and spring, being absent in summer, when insolation is strongest, due to the surface being snow-free in both the control and enhanced CO_2 climates. Consequently the increase in annual mean temperature upon increasing CO_2 was smaller in the seasonal model.

An interesting feature of Manabe and Stouffer's model, which contrasts with the results of Ramanathan et al.'s (1979) energy balance model described in Chapter 6, is that maximum warming occurred in high latitudes (especially in the northern hemisphere) in early winter. This was due to a much larger heat flux to the atmosphere from the ocean through a thinner ice layer. In summer when ice cover was substantially reduced or non-existent, the additional heating due to increased carbon dioxide was taken up by the ocean. The stored heat delayed the appearance of sea ice in autumn and reduced its thickness the following winter, and a large flux of heat from the ocean in autumn reduced substantially the usual high latitude seasonal cooling. This is clearly evident in the change in surface temperature in high latitudes of the northern hemisphere in early winter shown in Fig. 24. Over the continents the snow-albedo feedback produced a secondary maximum warming in spring. In the southern hemisphere the smaller land surface area rendered the snow-albedo feedback smaller than in the northern hemisphere. Also the temperature was insufficient to produce an ice-albedo feedback over Antarctica, so the overall warming there was less than in the Arctic.

It should be noted that in a doubled-CO_2 experiment, as opposed to the quadrupled-CO_2 experiment discussed here, the high latitude warming would probably be relatively less intense since substantially smaller areas of sea-ice would be expected to melt in summer. It is very clear that a proper treatment of the thermal properties of the ocean and the behaviour of sea-ice are essential for a reliable assessment of the seasonal variation of CO_2-induced warming in high latitudes.

Both the GISS models incorporated realistic, global geography and predictive cloud schemes. One employed a mixed layer ocean with seasonal variations, the other a swamp model with no seasonal forcing. These models gave global, annual average warming for a doubling of carbon dioxide of 3.5 K and 3.9 K, respectively. As described earlier, differences in the cloud prediction schemes are probably responsible in part for the higher value found in the GISS models compared with the GFDL models. The GFDL and GISS simulations also differed in their modelling of snow and ice cover, the snow-ice boundary extending much further toward the equator in the GISS than in the GFDL non-seasonal models. This led, in the opinion of the Ad Hoc Study Group, to an overestimate of the ice-albedo feedback in the GISS experiment and an underestimate in the GFDL experiment. The different positioning of the ice boundary also led to differences in the high latitude seasonal response to increased carbon dioxide between the GISS and GFDL seasonally varying models. The substantial melting of ice in summer and consequent ocean thermal storage noted in the CO_2-quadrupling experiment of Manabe and Stouffer was not an evident feature of the equivalent GISS simulation using doubled carbon dioxide concentration. In this case the effects of ice-thinning were most apparent in spring.

A summary of the characteristics of the models dealt with above is given in Table 4, which also gives the global mean and 70° latitude temperature increases obtained by each model for a doubling of the CO_2 concentration.

CO_2-induced changes in the water budget

The above discussion has concentrated on the latitudinal and seasonal variations of CO_2-induced temperature changes. Recently attention has been focussed upon the changes in the water budget which might arise as a result of the temperature and circulation changes associated with increased CO_2 concentrations and which may be of greater practical importance than changes in temperature *per se*.

It was noted earlier that increased poleward transport of latent heat due to the overall warming and moistening of the atmosphere contributed to greater warming at high than at low latitudes. In Manabe and Wetherald's (1980) annual mean sector model this led to a marked increase in precipitation (P) north of 50°N which exceeded the increases in evaporation (E) and gave rise to increased water availability (Fig. 25). Between 40 and 50°N increased CO_2 led to a reduction in (P-E) and a consequent reduction in soil moisture. Comparing models with and without the seasonal cycle (Wetherald and Manabe, 1981) showed (Fig. 26) that the annual mean difference in zonal mean (P-E) between present and quadrupled CO_2 integrations was less latitudinally variable in the seasonal model. As shown in Fig. 27, this was due largely to the seasonal excursion between about 35 and 55°N of the middle latitude zone of enhanced aridity. The seasonal model further indicated that the largest decreases in soil moisture occurred in summer, both in mid-latitudes and also in high latitudes, the latter feature being masked in the annual mean value by large increases in soil moisture in winter. The cause of the increased summer dryness was considered to be earlier and less copious snow-melt, greater evaporation in spring and early summer and an earlier onset of the spring to summer reduction in rainfall in the quadrupled-CO_2 case.

These experiments have been further advanced by Manabe et al. (1981) who compared results from the seasonal sector model just described with those from two similar models

Model Characteristics	MW75	MW80	MS80	$WM81_1$	$WM81_2$	H1	H2
Domain	$0° < \lambda < 120°$ $0° < \phi < 81.7°$	$0° < \lambda < 120°$ $0° < \phi < 90°$	Global	$0° < \lambda < 120°$ $-90° < \phi < 90°$		Global	Global
Land-ocean distribution	Ocean for $60° < \lambda < 120°$ $0° < \phi < 66.5°$	Ocean for $60° < \lambda < 120°$ $0° < \phi < 90°$	Realistic	Ocean for $60° < \lambda < 120°$ $0° < \phi < 90°$		Realistic	Realistic
Ocean	Swamp	Swamp	Mixed layer	Mixed layer		Mixed layer	Swamp
Seasonal variation	No	No	Yes	Yes	No	Yes	No
Cloud feedback	No	Yes	No	No		Yes	Yes
Horizontal resolution	About 500 km on a Mercator projection	5° in long. 4.5° in lat.	Spectral models with max. zonal wavenumber 15	Spectral models with max. zonal wavenumber 15		10° in long. 8° in lat.	10° in long. 8° in lat
Global mean surface temp. rise (K)	2.9	3.0	2.0*	2.4*	3.0*	3.5	3.9
Surface temp. rise at lat. 70°	~7	~6	~4*	~4*	~6*	?	?

Table 4: Summary of characteristics of GCMs mentioned in text. Also given are the global mean temperature rise (K) and the zonal mean temperature rise at 70° latitude (mean of 70°N and 70°S where appropriate) for a doubling of CO_2 concentration. An asterisk implies that the value given is half the value obtained in a CO_2-quadrupling experiment. H1 and H2 are the models of Hansen et al. at GISS. The remaining models are those of GFDL reported by Manabe and Wetherald (1975) - MW75; Manabe and Wetherald (1980) - MW80; Manabe and Stouffer (1979 and 1980) - MS80, Wetherald and Manabe (1981) - $WM81_1$ and $WM81_2$. The latitudinal distribution of the rise in temperature in the GISS models is unavailable.

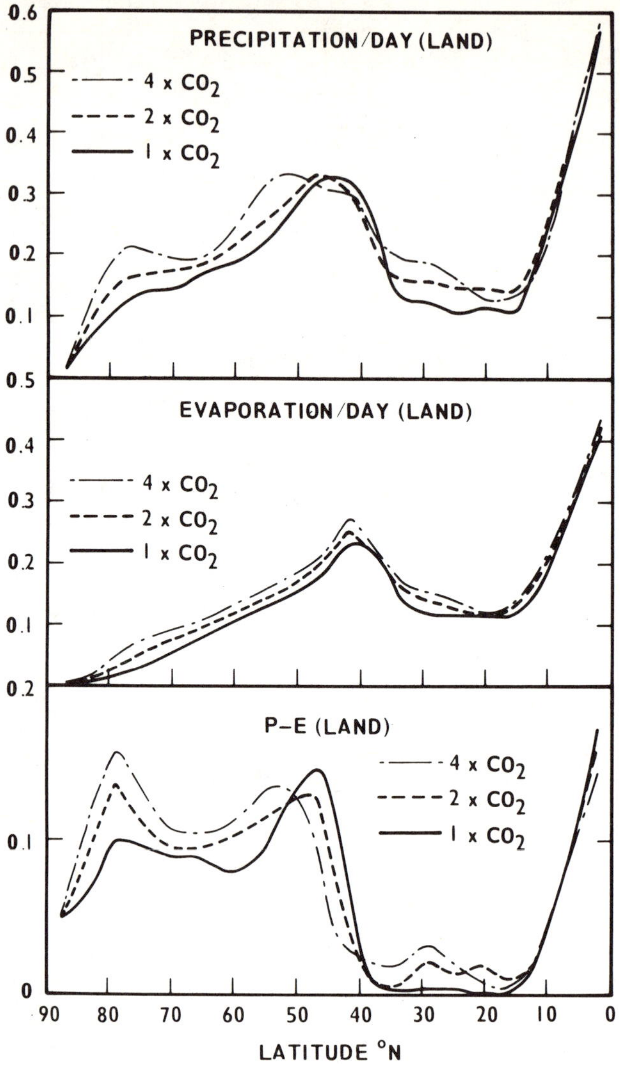

Figure 25. Latitudinal distribution of zonal-mean values of (a) precipitation rate, P, (b) evaporation rate, E and (c) (P-E) over the continent for present, doubled and quadrupled CO_2 contents in the Manabe and Wetherald (1980) annual mean sector model. Units are cm day^{-1}. (Reproduced from Manabe and Wetherald, 1980)

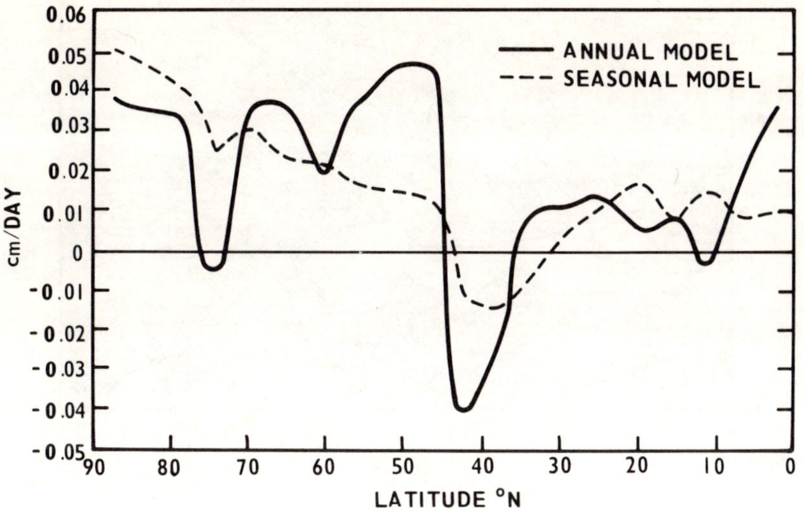

Figure 26. Latitudinal distribution of the differences in zonal mean (P-E) over the continent between present and quadrupled CO_2 content for annual mean (solid line) and seasonal (dashed line) versions of the Wetherald and Manabe (1981) sector (mixed layer) model. The distribution from the seasonal model shows the annual mean values. (Reproduced from Wetherald and Manabe, 1981)

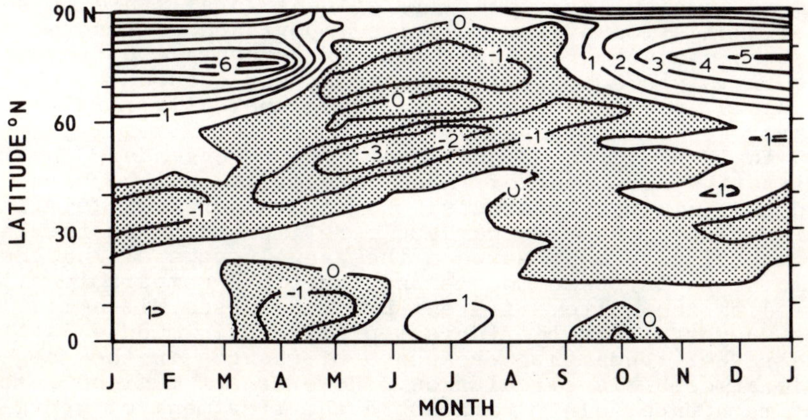

Figure 27. The latitudinal and seasonal variation of zonal mean soil moisture difference (cm) between present and quadrupled CO_2 contents for Wetherald and Manabe's (1981) sector model. (Reproduced from Wetherald and Manabe, 1981)

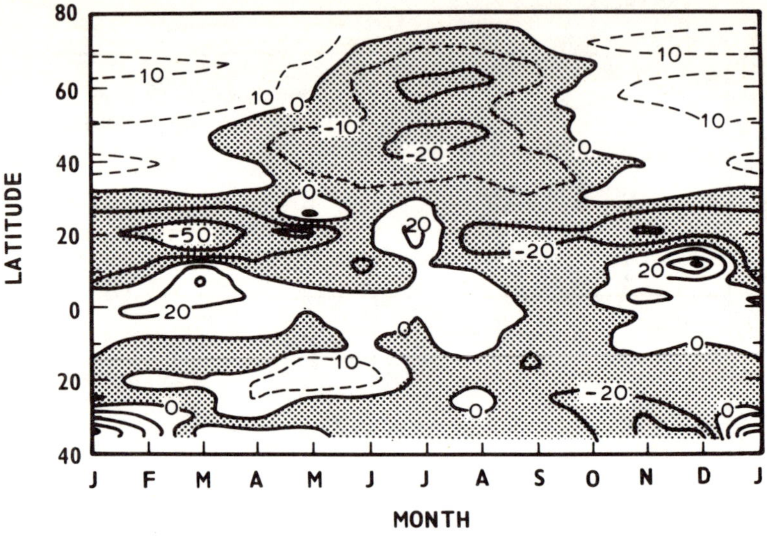

Figure 28. Percentage change of soil moisture from present to quadrupled CO_2 content. The average values from the two geographically realistic models are shown. Values polewards of 75°N and 35°S are not shown due to small land area. (Reproduced from Manabe et al., 1981)

containing realistic geography, one having a higher horizontal resolution than the other two models. While the changes were smaller in the models with realistic geography, the summer dryness in middle and high latitudes and increased winter wetness in high latitudes were evident in all three cases. As an indication of the relative importance of the soil moisture changes, Fig. 28 shows the percentage difference in soil moisture between the present and quadrupled CO_2 experiments, averaged for the two geographically realistic models.

While a degree of confidence may be engendered by the qualitative agreement of these models, it must be borne in mind that, in many respects, the models were very similar or identical. What is perhaps more impressive is the large measure of agreement between the annual cycle of the changes in moisture in these models and that in a comparable simulation in the quite different BMO model (see Mitchell, 1983). This suggests that the increased dryness in middle latitudes is a rather fundamental response of changes in the large scale atmospheric circulation. However, it must be remembered that much uncertainty remains in the treatment of hydrological processes as indicated in earlier discussions of humidity profile parametrizations and representation of clouds. Thus, reliable estimates of the impact of CO_2 increases on the latitudinal and seasonal distribution of hydrological parameters must await further experimentation. The results to date are probably best considered as having contributed to our understanding of climatic interactions which are likely to be important in CO_2-induced perturbations of the hydrological cycle.

Many of the published reports of these and other model experiments involving realistic geography have included details of the geographical changes in model climate as a result of increasing CO_2. Discussion of these has been omitted in this review so that any temptation to misinterpret the results as 'regional forecasts' can be avoided. The climate modellers are keen to stress that the models are not yet sufficiently advanced or reliable for use in regional climate sensitivity studies, reasons for which have already been discussed.

The transient response to increasing CO_2 concentrations

The climatic changes indicated by the experiments described so far in this chapter are those which might be expected if the CO_2 concentration were suddenly doubled (or quadrupled) and the system allowed to come to equilibrium. As computer and modelling capabilities advance it will become possible to simulate the time-dependent climatic change which occurs as the CO_2 concentration increases. Apart from its intrinsic importance, a knowledge of the likely transient response will be valuable in developing strategies for detecting CO_2-induced climatic changes.

An important factor in the transient response will be the effect of the ocean's thermal capacity in retarding the rise in temperature. First estimates of the lag caused by oceanic heat storage were made using simple, globally averaged models. Hunt and Wells (1979) showed that coupling a 300 m deep mixed layer ocean to a one-dimensional radiative-convective model introduced a lag of about 8 years in the surface temperature response to increasing CO_2, compared with the same model without an ocean. A delay of about two decades was estimated by Cess and Goldenberg (1981) using a global surface energy balance model coupled to a one-dimensional box-diffusion ocean model. In similar experiments values for the global mean lag between 5 and 20 years were obtained by Thompson and Schneider (1979) and Hoffert et al. (1980).

The possibility of latitudinal variations in the lag has been studied by Schneider and Thompson (1981). They noted that owing to the latitudinal dependence of the thermal inertia and energy exchange mechanisms, the approach towards equilibrium of the surface temperature of different regions of the globe will differ significantly from the global mean approach. Such variability may introduce climatic changes different from those suggested by the equilibrium simulations.

A first attempt at a fully synchronous atmosphere-ocean GCM simulation has recently been reported by Bryan et al. (1982). They described the model's time-dependent response to a step doubling of CO_2. Initially, asynchronous coupling was employed to determine the equilibrium model climates for present and quadrupled CO_2 concentrations. Then, in a synchronous integration starting from the equilibrium climate for present CO_2 concentration, the CO_2 was suddenly quadrupled and the time dependent approach of the model towards the equilibrium quadrupled-CO_2 climate determined previously was simulated. Fig. 29 illustrates the zonal mean surface temperature response, the values plotted representing the

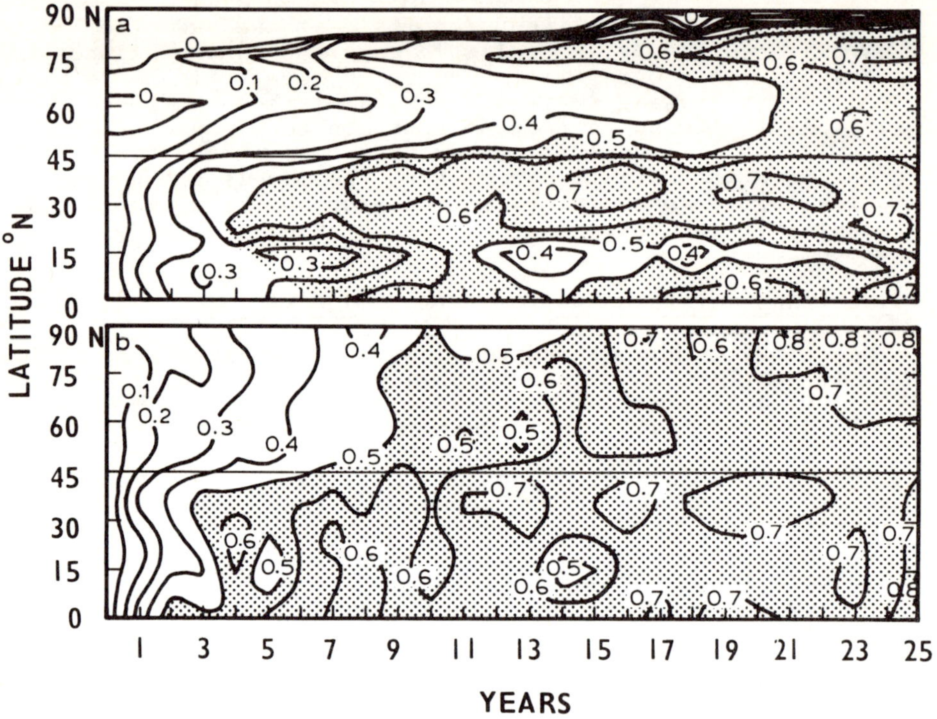

Figure 29. Time-latitude variation of the zonal mean normalised response (see text) of (a) sea surface temperature and (b) surface air temperature (averaged over continent and ocean) to a step-function quadrupling of CO_2 for the synchronously coupled atmosphere-ocean model of Bryan et al. (1982). (Reproduced from Bryan et al., 1982, Copyright 1982 The American Association for the Advancement of Science)

change in temperature as a fraction of the difference between the previously modelled present and quadrupled-CO_2 climates. Fig. 30 shows the zonally averaged temperature increase after 10 years as a function of latitude and atmospheric height/oceanic depth. An important feature is the relatively rapid rise in sea surface temperature in the tropics, where vertical mixing in the ocean is relatively weak, and the much slower response in the higher latitude, deep mixing regions. While the latitudinal variation in the response might seem to support the views of Schneider and Thompson, Bryan et al. argued that with the much less extreme perturbation of the system associated with gradually increasing CO_2 concentration, the approach to equilibrium would be much more uniform with latitude as indicated by the response towards the end of their integration. It should be noted that the quantitative aspects of the results are not to be relied upon, since to permit a 25-year simulation, idealized 'sector' geography and neglect of seasonal variation were necessary. The qualitative conclusion drawn by Bryan et al. is that the oceans can delay, but by no means eliminate, a strong climatic response to increasing atmospheric CO_2.

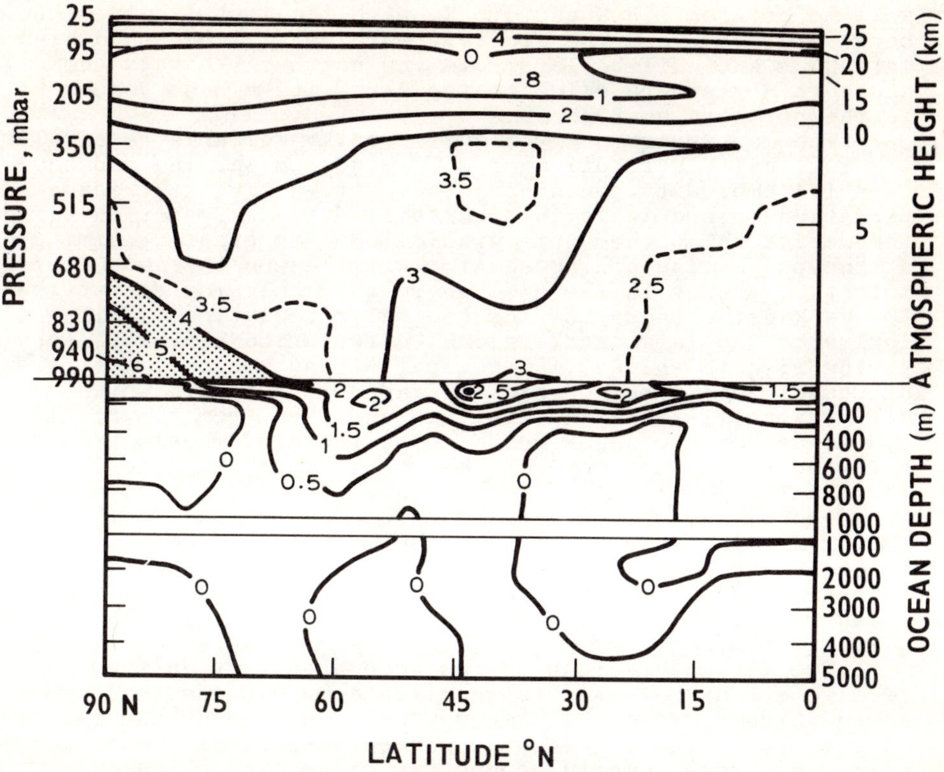

Figure 30. Latitude-height distribution of the departure of zonal mean temperature (K) at 10 years from the initial temperature due to a step-function quadrupling of CO_2 at year zero for the same model as in Fig. 29. (Reproduced from Bryan et al., 1982, copyright 1982 The American Association for the Advancement of Science)

<u>Summary of GCM CO_2 studies</u>

Mainly due to lack of available computing power and the expense of running large scale climate models, none of the models discussed above has been able to incorporate all the physical processes thought to be important in influencing the change in climate arising from a doubling of carbon dioxide concentrations. The results of those models believed to be most realistic show a globally averaged warming of between 2 and 3.5 K. On the basis of their examination of the problem, the NRC Ad Hoc Study Group (National Academy of Sciences, 1979) concluded that 'the oversimplifications and inaccuracies in the models are not likely to have vitiated the principal conclusion that there will be an appreciable warming'. The results from both the radiative-convective and general circulation models suggests that the equilibrium surface global warming due to doubled carbon dioxide, when reasonable errors of estimates are included, will be between 1.5 K and 4.5 K, with the most likely value between 2 and 3 K. The greatest warming will occur in high latitudes and will be larger in the northern than in the southern hemisphere owing to the large ice-albedo feedback in the northern hemisphere.

The models are not yet sufficiently reliable to simulate the regional distribution of a CO_2-induced warming and there is still ambiguity concerning the pattern of likely seasonal variations, especially in high latitudes. A reduction in the meridional temperature gradient due to greater warming at the poles than at the equator will lessen the amount of potential energy in the atmosphere available for conversion to the kinetic energy of the general circulation. This will influence the large scale global circulation patterns, thus giving rise to changes in the spatial and seasonal distributions of temperature, rainfall and soil moisture which may differ significantly in magnitude and sign from the zonal and annual mean changes. The comparisons of models presented have indicated that some of the results are very sensitive, even in the global mean, to the ways in which atmosphere-ocean coupling, clouds and ice and snow are modelled. Improvements in these aspects of the models and the inclusion of ocean currents will be required before reliable regional climatic changes can be predicted. Finer horizontal resolution will be necessary for the modelling of certain regional climatic features.

Two important areas of research have only just begun to receive attention. Realistic climate models must be developed which address the time-dependent nature of the increase in atmospheric carbon dioxide concentration. These will require much more powerful computers on account of the very long periods of integration necessary (50-100 years). The delaying role of the deep ocean's thermal capacity will need to be modelled. The results of these experiments will of course depend on the accuracy with which we can predict the future increases in the atmospheric carbon dioxide concentrations (see page 93).

A second requirement of the climate research effort is the prediction of the variability of future climate, which may differ from that experienced at present. Knowledge of a change in the probability of occurrence of extreme climatic conditions may be as important to agriculturalists, for example, as knowledge of a change in the mean conditions. Integrations of many model years are required for the evaluation of the statistics of the natural variability of model climates. In consequence, not much work has yet been done in validating the variability of model simulations of present-day climate against climatological statistics. Not only is a knowledge of this variability important in itself, but it is also necessary to enable proper judgements to be made of the significance of climatic changes which are predicted in response to a change in model forcing, such as a doubling of carbon dioxide.

9 Clues from the Past

Warm world scenarios based on historical analogues

The previous discussion has indicated that numerical models of climate are as yet unable to predict the likely regional variations in climate change induced by increasing carbon dioxide concentrations. In order to obtain an idea of the possible magnitude of such changes, several studies have been made on climatic data from periods when the climate was generally warmer than at present. Since the causes of previously warm periods may have been factors other than increased carbon dioxide, the authors of these researches stress that the results must be regarded as indications of the scale of the changes which could occur, not as geographical predictions of future climatic change. However, it is possible that similar regional climatic changes could result from very different global climatic change mechanisms. Some tentative support for this view is found in the model of Manabe and Wetherald (1980) which responded very similarly to two different climatic forcings: one a doubling of carbon dioxide, the other an increase of 2% in the solar constant. This was ascribed to the fact that the influence of feedback processes, triggered by the general warming in both simulations, overwhelmed the differences between the latitudinal distributions of the two radiative forcings in the determination of the new meridional temperature gradient. If this apparent insensitivity of the pattern of climatic change to the forcing mechanism is a real feature of the system, the relevance of this line of research to the CO_2 issue may, in principle, be greater than some believe.

Kellogg (1977) has examined evidence from the hypsithermal period between 4000 and 8000 years ago when the northern hemisphere was warmer than now. During that period summertime rainfall over large areas of Europe, north and east Africa, India, China and Alaska was greater than today, while much of Canada and the United States was drier. However, in view of the relatively long period over which the data are probably spread, it is possible that the various regional differences did not all occur at the same time.

Since the high latitudes are likely to experience greater warming than the tropics, Williams (1980) has examined the large scale circulation anomalies associated with the 10 warmest Arctic winters and the 10 warmest Arctic summers during the last 70 years. The work demonstrated that large, coherent anomalies of temperature occur elsewhere in the hemisphere when the Arctic is warm, and that the anomalies are not necessarily all positive. As a result of circulation changes, large coherent anomalies in rainfall distribution also occur. The magnitudes and distribution of the anomalies vary with season. For the years studied, some, but not all, anomalies were found to be statistically significant. Wigley et al. (1980) followed a similar approach, comparing annual mean conditions of the five warmest years with those of the five coldest years in the zone 65° to 80°N during 1925-74. Comparison of these analogue studies reveals that different patterns of temperature anomalies occurred. However it may be possible from an increased number of such studies to identify certain features of regional climatic change that tend to occur repeatedly in a variety of assumed warm-world analogues and which therefore may be expected to occur in a climate warmed by increased carbon dioxide concentrations. Certainly research of this nature has the potential to complement the numerical simulations of regional climatic change that should be possible in the future.

Past atmospheric CO_2 concentration deduced from ice cores

Reconstruction of the atmospheric CO_2 content in previous centuries and millennia can, in principle, be determined by analysis of the CO_2 content of air trapped in natural ice. Relating past CO_2 levels to the then prevailing climates should provide valuable information on the interaction between CO_2 and climate, although cause and effect relationships may be difficult to deduce. Taking the analysis a stage further, knowledge of the isotopic composition of historical atmospheric CO_2 could assist understanding of the mechanisms within the global carbon cycle leading to CO_2 changes.

The experimental difficulties associated with the extraction of uncontaminated air samples from bubbles trapped in ice cores have, until very recently, posed insurmountable problems. Techniques in which the ice samples were melted tended to yield extracted air rich in CO_2 due to the natural or artificial (by contamination) presence of carbonates in the ice lattice (Delmas et al., 1980). New methods, involving either the separation of the CO_2 in the air bubbles from that released from the ice lattice (Berner et al., 1980) or dry extraction (Delmas et al., 1980; Neftel et al., 1982), have yielded more reliable measurements of past CO_2 levels, although the results must still be treated with caution.

An estimate of the atmospheric CO_2 concentration over the past 40 000 years, deduced by Neftel et al. (1982) from ice samples from Greenland and Antarctica, is shown in Fig. 31. This result is in good agreement with those of Delmas et al. (1980) based on two different Antarctic cores. The minimum concentrations of about 200 ppmv coincide with the peak of the last glaciation between 15 000 and 20 000 years ago, with an increase occurring during the climatic recovery towards the warm, hypsithermal period some 8000 years ago.

It is, of course, tempting to ascribe the climatic warming to the increasing levels of CO_2, but it is not clear if the CO_2 increase preceded or lagged the temperature change. The global mean temperature increase over this period is reckoned to have been well over 2 K which, on the basis of climate model results, is considerably more than would be expected from an increase in CO_2 concentration of around 30-40% suggested by the ice core results. However the removal of a greater extent of ice could have led to a much stronger ice-albedo feedback than would occur today and consequently a larger warming for a given CO_2 increase could have resulted.

It is probable that the increase in temperature was caused, in part, by increasing CO_2 but was itself also a cause of increased CO_2. Various explanations have been suggested for the rise in CO_2 levels. Berner et al. (1980) point out that sea surface temperature changes alone could not account for the difference between the CO_2 concentrations in the glacial and warm epochs. Rather, mechanisms for altering the total carbon content of the mixed layer are the most likely candidates since a change of 1% in the surface ocean total carbon content leads to a change of about 10% (the buffer factor) in the dissolved CO_2 partial pressure and hence in the atmospheric concentration (see Chapter 2). The carbon contents of surface waters are depleted relative to deeper layers as a result of biological productivity.

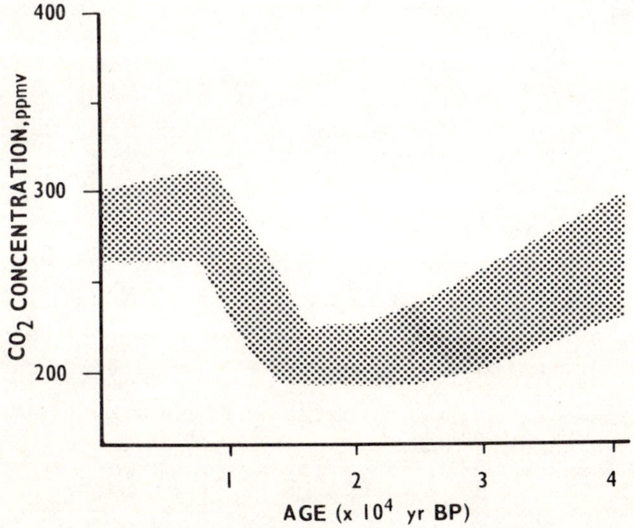

Figure 31. Range of atmospheric CO_2 during the past 40 000 years estimated from ice core analysis. (Reproduced from Neftel et al., 1982, reprinted with permission from Nature, copyright 1982, Macmillian Journals Ltd.)

Thus any reduction in biological activity as the glacial-interglacial transition proceeded could have led to higher surface water carbon concentration and an increase in the atmospheric CO_2 loading. Broecker (1980b) has postulated that a reduction in the availability of phosphorus could account for a decline in productivity and that a rise in sea level as the ice sheets melted could have led to phosphorus depletion.

While present estimates of past CO_2 appear to fit our knowledge of past climate reasonably well, it must be borne in mind that they have been calculated from only a few ice cores of very limited geographical distribution and amidst considerable experimental difficulties. There is much potential for development of these techniques to yield more accurate and definitive records of past CO_2 concentration. At present the results and theories to explain them must be considered tentative.

PART III

Implications of Increased Carbon Dioxide for the Use of Energy in the Future

Introduction

So far in this book we have examined the strength of the relationship between increase in atmospheric carbon dioxide and climatic change. Most studies seem to indicate that a doubling of CO_2 in the atmosphere will lead to a mean temperature increase in the lower troposphere of 2-3°C; the time required to produce such a doubling will be discussed shortly. A man-made shift in climate of this magnitude would clearly be significant since we have to go back at least 10^4 years in order to find changes of this size occurring naturally (Wigley, 1982). A discussion of the implications of such a warming for society in general is beyond the scope of this review; they have been assessed in detail by Kellogg and Schware (1981). However, there are a number of questions which are of particular concern to the energy industries. These include:

(i) What would be the direct effect of such a climatic shift on energy demand?

(ii) What is the timescale of the increase in atmospheric carbon dioxide?

(iii) When will we know with reasonable certainty that the temperature of the Earth's surface is really going to experience a significant CO_2-induced warming?

(iv) Are there any technological solutions which might be used to reduce (or halt) the rate of atmospheric CO_2 increase?

(v) What role should individual nations and their energy industries play in research into the CO_2/Climate issue nationally and internationally?

These various questions are discussed in turn.

The direct effect of climatic change on energy demand

The way in which a CO_2-induced climatic change would influence energy demand would vary across the globe, due both to the likely regional character of the climatic change and to different patterns of energy use. In the following we use figures appropriate to energy use in the UK as an example and assume the global mean warming.

a) Electricity

Some indication of the likely magnitude of the effect of a climatic change can be obtained from current UK experience of the change in demand for electricity as temperature varies with day-to-day alterations in the weather. Clearly such information cannot be used to assess accurately changes in demand due to climatic changes, which are by definition long-term, since there would presumably be some adaptation to cope with the changing temperature regime. In spite of this uncertainty, extrapolation of present experience from short-term temperature variations is probably useful if order of magnitude estimates only are required.

The sensitivity of present electricity demand to temperature, averaged over the whole year, is about 1% per degree centigrade, i.e. if the temperature is 1^oC lower than average then the demand for electricity increases by approximately 1% over that normally used. Yearly averaging covers up seasonal variation in the sensitivity which is greatest in spring and autumn (about 2% $^oC^{-1}$), very small in summer, and near the yearly average in winter. Thus it would seem that even a 3^oC warming of the climate would lead to less than a 5% decrease in demand for electricity, even assuming no societal adaptation to the change in temperature, e.g. by installation of more air conditioning plant to cope with the warmer summers, which would not only tend, on a yearly basis, to back-off the decreased use of electricity for space and water heating, but would also lead to a lessening of the present differential use between winter and summer.

b) Oil and gas

The supply position of the UK oil and gas industries is more sensitive to the effect of temperature than is that of the Electricity Supply Industry. Thus, the temperature sensitivity of oil supplies in winter is between 4% (fuel oil) and 5% (gas oil) per ^oC temperature change (Tellings, 1978). Estimates from British Gas indicate a typical sensitivity of about 4% $^oC^{-1}$.

All the temperature sensitivities given in this section are expressed as percentage changes in the *total* amount of the particular fuel or energy supplied. Hence, the higher temperature sensitivities of oil and gas compared with electricity almost certainly arise because for the former much more of the amount supplied is used for space heating where demand is rather susceptible to outside temperature. For example only about 20% of the total amount of electricity generated in the UK goes into space (and water) heating, whereas for fuel oil and gas oil 45% and 61%, respectively, of the amounts sold are used for heating purposes. If one just considers energy demand for space heating alone, then temperature sensitivities as high as 8-9% $^{o}C^{-1}$ (Tellings, 1978) and 10% $^{o}C^{-1}$ (McKay and Allsopp, 1980) have been predicted.

The conclusion to be drawn seems to be that the direct effect of climatic change on the electricity supply industry is likely to be rather small, whereas it may be more significant for the oil and gas industries, e.g. about a 12-15% decrease in total demand for a 3°C rise in mean temperature, with no compensating increase in amount supplied arising from possible greater use of air conditioning in buildings.

The time-scale of atmospheric CO_2 increase

In order to calculate the timescale of future increase in atmospheric CO_2 it is necessary to make several assumptions, all of which are open to question, so that the resulting predictions are similarly uncertain. For our calculation we start with the atmospheric concentration of CO_2 at 339 ppm in 1982. We then assume that in the future the mix of fossil fuel which is burned will remain as at present. Although, at first sight, this might not seem to be an important point, a dramatic change in fossil fuel mix could significantly alter the amount of CO_2 emitted. Table 5 shows the amounts of carbon (as CO_2) produced per unit of thermal energy for different fossil fuels. The relatively low CO_2 output from gas arises from the high proportion of the energy (~60%) derived by conversion of H atoms to H_2O. In contrast, for coal this process is much less significant since 80% of the energy comes from the conversion of C to CO_2. The much larger CO_2 output per unit of energy for synthetic fuels (more than a factor of 2 over that for gas) arises because energy has to be used, and CO_2 produced (assuming this energy comes from fossil fuels), in converting coal to synfuels (whether oil or gas). Clearly a major shift to synthetic fuel production would substantially increase CO_2 emissions; a point which was made forcibly in the United States over the proposals made several years ago for major investment in synfuel production facilities. Even without synfuels, replacing all the oil and gas currently used by coal would increase CO_2 emissions by approximately 25%. Our assumption of no change in fossil fuel mix in the future is probably conservative in CO_2 emission terms since, if anything, there is likely to be some shift towards coal (and maybe synfuels) and away from oil and gas. This arises because of the much greater known reserves of coal over those for oil and gas (Rotty and Marland, 1980, Table 5).

Another uncertainty in projections of future CO_2 levels is over what figure to take for the proportion of the amount of CO_2 emitted which will remain in the atmosphere. The historical apparent airborne fraction is about 0.55 but, as discussed on page 41, uncertainties over the role of the biosphere, mean that the true value could be as low as 0.38 or as high as 0.67. Until we better understand how anthropogenic CO_2 is partitioned between the various environmental

Table 5: Carbon Dioxide Produced Per Unit of Energy Generated from Different Fuel Types

Fuel	Carbon (as CO_2) produced per 100 Quads* of thermal energy (GtC)
Oil	2.0
Gas	1.45
Coal	2.5
Synfuel derived from coal	3.4

* 1 Quad = 10^{15} BTU (1.055×10^{15} kJ)

reservoirs, prediction of what values the airborne fraction will have in the future is clearly very difficult. In the projections given here we have used a value for the airborne fraction (α) of 0.55 and also repeat the calculation for the extrema of the historical ratio (0.38 and 0.67) suggested by Oeschger and Heimann (1981).

By far the greatest uncertainty in projections of future atmospheric CO_2 levels is what figure to assume for the growth rate of fossil fuel consumption. As discussed in Chapter 1 the historical rate of increase over the last century has been 4.3% $year^{-1}$. Since 1973 the rate has slackened to about 2.25% $year^{-1}$ (see Fig. 6). For present purposes growth rates (r) of 4%, 2% and zero have been adopted, to represent somewhat less than the historical and recent rates of increase, and the zero growth scenario.

The results of the calculation are shown in Fig. 32. For each of the three growth rates a family of curves is shown corresponding to the range of assumed airborne fractions. The overwhelming importance of fossil fuel growth rate for future atmospheric CO_2 is immediately apparent. This can be illustrated by reference to the year at which p_{CO_2} will reach 600 ppm (i.e. twice pre-industrial levels, as used in many of the climate models), as shown by the horizontal line in the figure. For r = 4% $year^{-1}$ the doubling is attained in 2036 (for $\alpha = 0.55$) with a range of from 2032-2045 for $\alpha = 0.67$ and 0.38, respectively. With r = 2% $year^{-1}$, 600 ppm is reached 24 years later in 2060 (range 2053-2076). For zero growth extrapolation of the lines shown in Fig. 32 indicates a delay in the doubling until well into the 22nd century, i.e. until 2172 (range 2138-2258). For atmospheric concentrations higher than 600 ppm the divergence between the three sets of curves is even more marked.

The projections given here indicate that unless fossil fuel growth rates are significantly below 2% $year^{-1}$ then sometime in the mid to latter part of the next century atmospheric CO_2 will stand at double its pre-industrial value. The uncertainties in this conclusion have been outlined above. It should be stressed that the calculation performed here

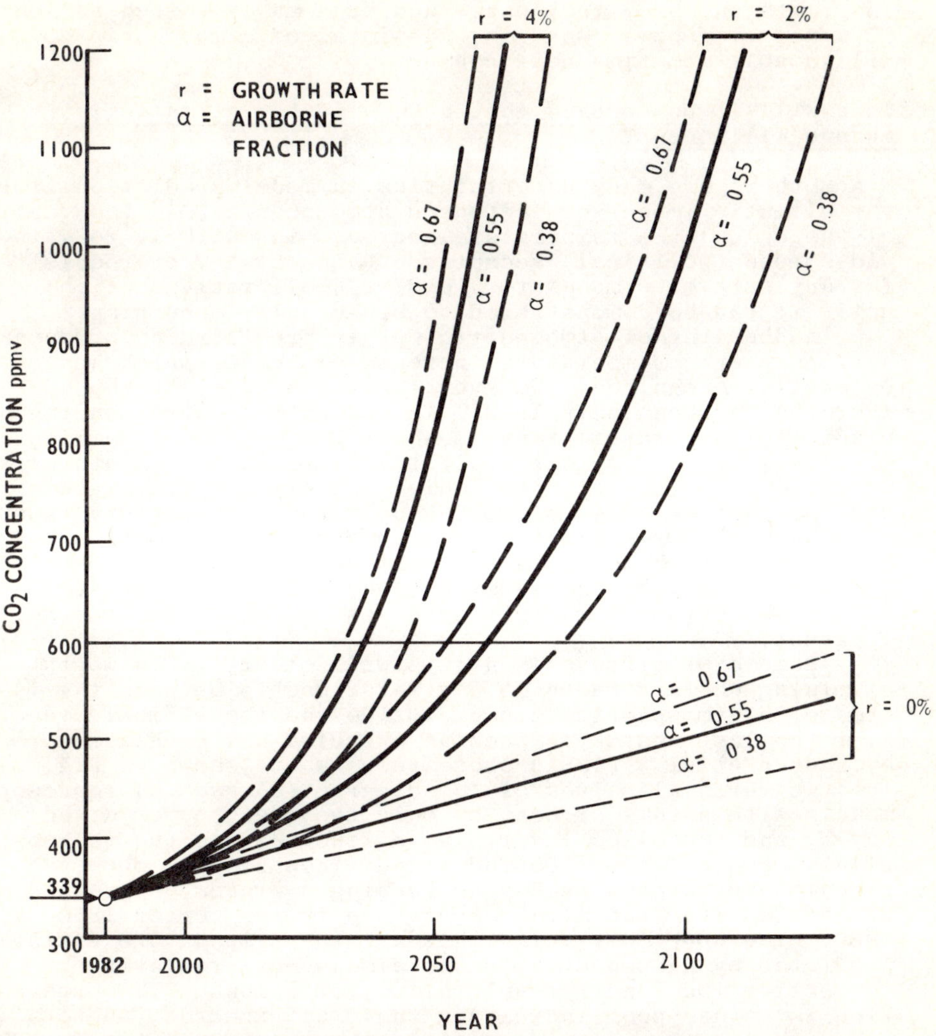

Figure 32. Predicted growth of atmospheric CO_2 concentration over the next 150 years, assuming growth rates (r) in CO_2 emissions of 0, 2 and 4% yr^{-1}, and airborne fractions of 0.38, 0.55 and 0.67

refers only to fossil fuel usage. A shift away from fossil fuel to energy sources which are not CO_2 producers (nuclear and renewables) would allow growth in energy output but with a lesser CO_2 penalty, i.e. the date at which a given level of atmospheric CO_2 would be attained would be delayed. Similarly, implementation of additional energy conservation measures would allow economic growth with a lower rate of increase in atmospheric CO_2. It is not our purpose here to present a detailed discussion of possible future changes in energy sources and conservation. However, it is abundantly clear from Fig. 32 that unless there is a dramatic decrease in growth rates of fuel usage and/or a major switch to non-CO_2 producing power sources, the level of atmospheric CO_2 will double sometime next century.

<u>When will it be certain that a CO_2-induced climatic change will occur?</u>

Since there are many uncertainties in model predictions for the climatic impact of increased atmospheric CO_2, the reality of the calculated temperature increase is unlikely to gain widespread (political) acceptance (a necessary prerequisite for any sort of compensatory and/or ameliorative action) until it can be demonstrated to be actually occurring.

A doubling of atmospheric CO_2 is predicted to produce a tropospheric temperature increase of 2-3°C, which would be easily recognised. However, up to the present the CO_2 increase has been only 15-25% for which the corresponding predicted temperature rise is about 0.2-0.3°C. A warming of this size is very difficult to see against the natural variability of the climate, which in terms of temperature has been of the order of $\pm$0.2°C over the past century. Assuming that the CO_2-climate models are accurate, sometime between now and whenever in the next century the atmospheric CO_2 level doubles, the warming trend will become greater than the noise in the climatic record, and so will become readily apparent.

Before we discuss when this will occur, it is worth examining whether there is any relationship between the increase in atmospheric carbon dioxide and the climatic record over the last century. Such an exercise has been attempted by Hansen et al. (1981) whose results are shown in Fig. 33. The two vertical halves of the diagram (a) and (b) represent models with uptake of heat by only the ocean upper mixed layer, and the mixed layer plus diffusion into the thermocline, respectively. Dotted trend lines are the observed global temperatures as 5-year running averages, expressed as deviations from the mean temperature in the period 1880-1980. The uppermost full lines are the temperature deviations predicted by Hansen et al.'s one-dimensional radiative-convective model as a result of CO_2 emissions. The prediction is rather poor indicating that, as expected, factors other than carbon dioxide can affect atmospheric temperatures. Hansen et al. use the Dust Veil Index (Lamb, 1970) to combine with the CO_2 effect the role of volcanic eruptions in increasing atmospheric particulates, which are well known to alter temperatures for several years after major eruptions. When this is done (middle curve Fig. 33) the fit between the observed and predicted temperatures is significantly improved. Finally, the effect of changes in solar luminosity is incorporated into the predictions, via observed variations in the umbra/penumbra ratio in sunspots, to obtain the lowest curves in the figure. Especially for the ocean model with uptake of heat by the thermocline, the predicted trend agrees remarkably well with the observations. This agreement would constitute strong evidence for the CO_2-model predictions if there was certainty over the other two factors. Unfortunately, neither the role of volcanic emissions nor, especially, the solar luminosity effect can be described with such certainty

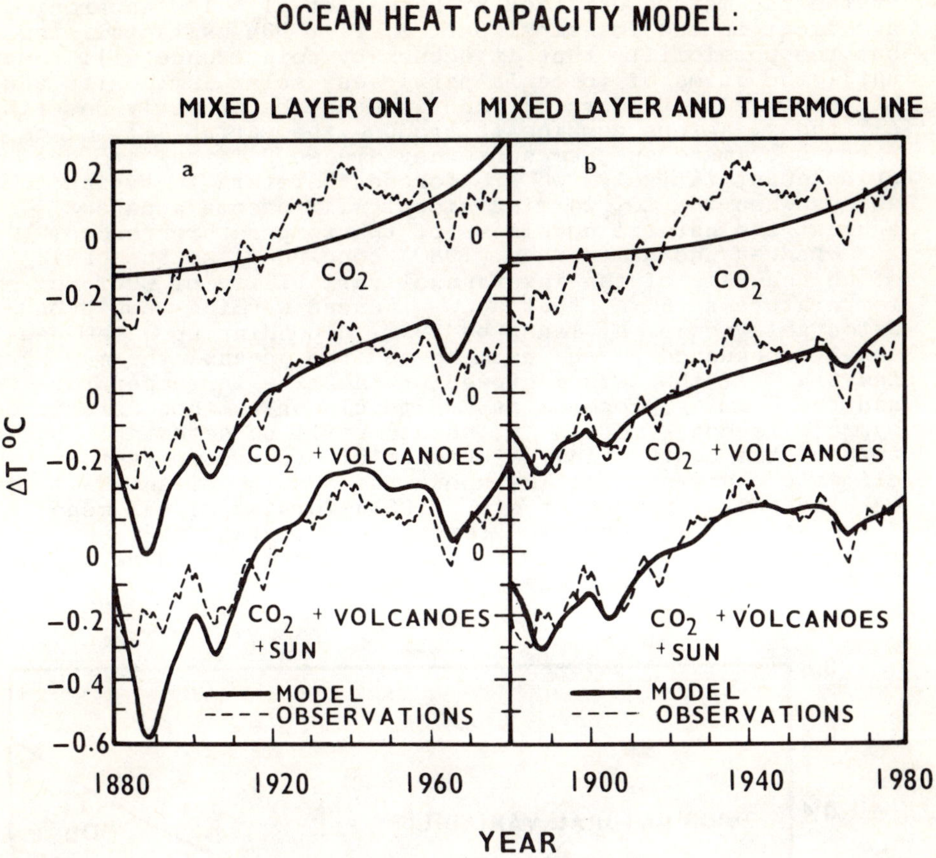

Figure 33. Global temperature trend (ΔT) obtained from a climate model with sensitivity +2.8°C for a doubling of atmospheric CO_2 (——), compared with observed temperatures (-----) over the period 1880-1980. The model results in (a) are based on a 100 m mixed-layer ocean for heat capacity, while those in (b) include diffusion of heat into the thermocline to a depth of 1000 m (coefficient of diffusion, $K = 1\ cm^2\ s^{-1}$). In the middle curves the effect of volcanoes is incorporated into the model predictions, while in the bottom curves both volcanic effects and changes in solar luminosity are included. The mean value of ΔT is zero for both the observations and the model results. (After Hansen et al., 1981, copyright 1981 The American Association for the Advancement of Science)

by the parameters utilized by Hansen et al. The apparent agreement in the lowermost part of Fig. 33b is tantalising, but the possibility that it occurs by coincidence will remain until the roles of volcanic particles, solar luminosity and any other climate forcing functions can be properly described. All these factors combine to produce the variability of the Earth's climate and, because they are so difficult to describe quantitatively, we are forced to return to trying to assess when the CO_2 warming signal will become apparent against the natural noisiness of the temperature record.

Madden and Ramanathan (1980) concluded, on the basis of an analysis of the interannual variability of surface temperature at 60°N, that a CO_2-induced warming should be detectable now or at least by 1990, depending upon the seasonal dependence of the warming and the oceans' thermal inertia. If the GCMs overpredict the rise in temperature and the warming proceeds at a rate closer to those predicted by zero-feedback models, it should still be detectable by the turn of the century in the absence of other compensatory climatic changes. Fig. 34 shows the results of an assessment carried out by Hansen et al. (1981) for the global mean temperature. The observed temperature deviations from the mean are shown together with one and two standard deviations of this variation projected 20 years into the next century.

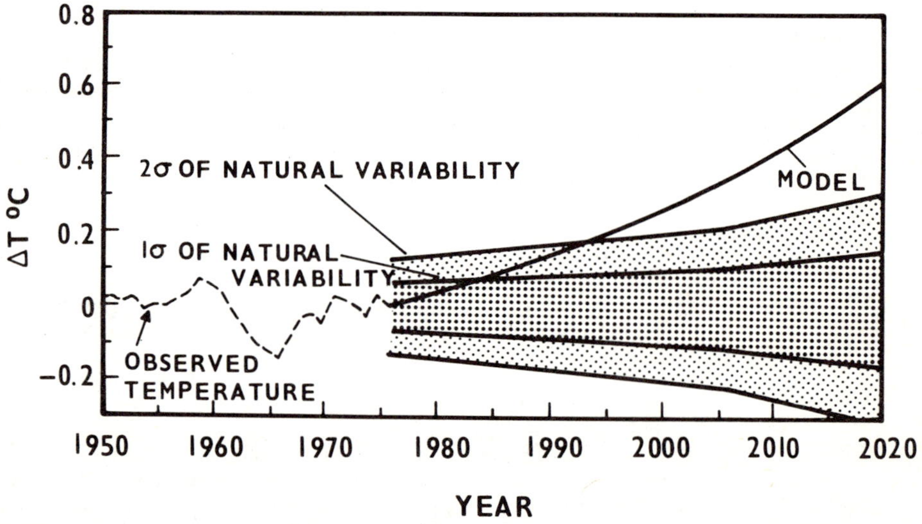

Figure 34. CO_2 warming versus noise level of natural climate variability. Comparison of CO_2 warming predicted from a climate model with sensitivity +2.8°C for a doubling of atmospheric CO_2 with one and two standard deviations (σ) of the observed global temperature record. (After Hansen et al., 1981, copyright 1981 The American Association for the Advancement of Science)

The temperature deviation produced by increased CO_2 as predicted by their 1-D model is shown and crosses the 1σ (85% confidence limit) and 2σ (98% confidence limit) lines in the mid 1980s and mid 1990s, respectively. Likewise, from a similar analysis, Wigley (1982) concludes that the CO_2 temperature signal will emerge from the noise sometime between 1990 and early next century. These predictions are not significantly affected by choice of fuel-use scenario since, on this short time-scale, deviations between different fossil fuel growth rates are not yet apparent (see Fig. 32).

Carbon dioxide is not the only infrared-absorbing trace gas with a long residence time which man is adding to the atmosphere in large enough quantities to yield significant increases in concentrations. Ozone, chlorofluorocarbons (CFCs), methane and nitrous oxide all have absorption bands in the Earth's infrared emission spectrum and thus contribute to the greenhouse effect (see Fig. 35). Carbon monoxide, nitric oxide and methane also contribute indirectly by enhancing the production of ozone in the troposphere (Ramanathan, 1980). Since methane, carbon monoxide and the oxides of nitrogen are released during fossil fuel combustion, their atmospheric concentrations might be expected to rise along with that of CO_2. The influence of these gases should therefore be taken into account when considering the timing of any induced climatic warming.

Ramanathan (1980) has demonstrated, using a vertical-column, radiative-photochemical model, that given reasonable future emission scenarios, increases in the trace gases other than CO_2 could contribute roughly 40% to the combined greenhouse warming of CO_2 and these gases. Three quarters of the temperature increase due to the gases other than CO_2 is not direct but results from the photochemical conversion into radiatively-active species of non-radiatively-active trace gas emissions. A similar study to estimate the equilibrium greenhouse warming from measured increases in trace gases in the decade 1970-1980 has been performed by Lacis et al. (1981). Their results, which compare well with those of Ramanathan (1980), are illustrated in Fig. 36. On this basis they predicted that the combined greenhouse warming during the 1970-1990 period should substantially exceed the natural variability of a 20 year period.

While it should be borne in mind that knowledge of tropospheric photochemistry involving these trace species is far from complete and that forecasts of future abundances are thus difficult to make, it is clear that increases in their concentrations will most likely bring forward the time at which a climatic warming is discernible and hasten the onset of any major changes.

It is perhaps appropriate to note here that increases in man-made aerosol concentrations in the troposphere would probably enhance the warming on a regional basis (Kellogg, 1980). The oft-quoted view that corresponding aerosol increases will offset a CO_2-induced warming are probably based upon experience relating to increased stratospheric aerosol loading following volcanic eruptions, in which case surface cooling occurs. The short residence time of several days, and hence predominance over land, and confinement to the troposphere of most anthropogenic aerosols gives rise to a

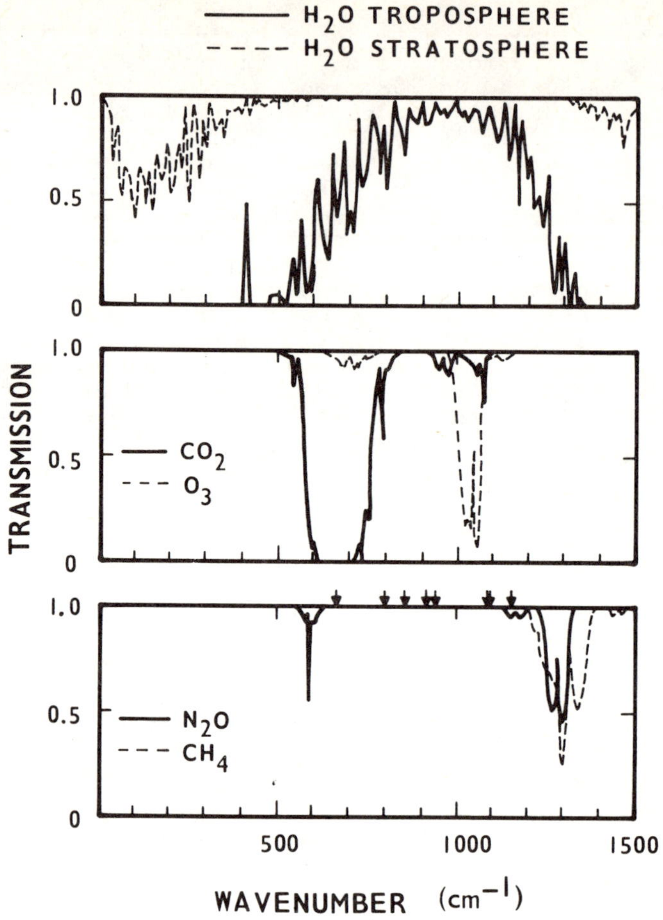

Figure 35. Fractional transmission of thermal radiation by atmospheric gases for present day abundances. Arrows indicate the locations of chlorofluorocarbon bands. (Reproduced from Lacis et al., 1981)

very different and more complex problem, to which knowledge of the effects of stratospheric aerosol increases cannot be applied directly.

The conclusion to be drawn from the preceding paragraphs is that by the turn of the century, and maybe earlier, it should, in principle, be possible to verify or reject by observation the climate model predictions. In practice, verification, or otherwise, will depend on the development of adequate strategies to enable early and confident detection of a CO_2-induced warming. Simultaneous identification of several of the different predicted effects on climate of increased CO_2 would be a strong indication of its being the cause. As noted in Chapter 8, further developments in modelling the transient response to actual CO_2 increases should

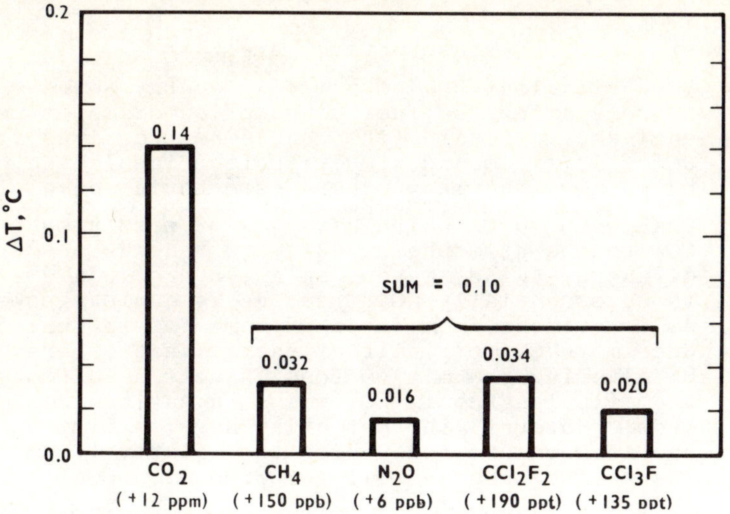

Figure 36. Equilibrium greenhouse warmings for estimated 1970-1980 abundance increases of several trace gases based on a one-dimensional radiative-convective model with sensitivity ~3°C for doubled CO_2. (Reproduced from Lacis et al., 1981)

point to those variables in the climate system which would most readily reveal CO_2-induced changes and to those areas of the globe where they should first occur. Once, however, a climatic change is confirmed with sufficient reliability for widespread acceptance of its existance, mankind will at that point have to decide what, if anything, should be done about it.

If the model predictions are confirmed sometime near the turn of the century, this would leave about 45 or 70 years (for 4% or 2% year^{-1} fossil fuel growth rates, respectively, and allowing a lag of 10 years to attain equilibrium warming) before the occurrence of the really substantial climatic warming predicted for a doubling of atmospheric CO_2 (2-3°C on average, but up to a maximum of about 8°C in polar regions). Although in the time available it would probably be technically feasible, using some or all of the options discussed next, to do quite a lot to prevent (or counteract) this climatic change, whether the international agreement necessary for such action could be achieved is an open question. It would be a problem second only to the abolition of war and roughly on par with the control of human population growth, and a discussion of this aspect is clearly beyond the scope of this book!

Technological options for reducing the rate of increase in atmospheric CO_2

Various possible measures exist which would allow energy production and use in the future but with a lesser penalty in terms of CO_2 emitted to the atmosphere than has occurred in the past. Some of these options are:

(i) More efficient use of existing energy sources e.g. by decreasing the fuel consumption of motor vehicle engines, increasing the usable heat derived from domestic fires and stoves, better insulation of buildings, and many others (see Lovins et al., 1982).

(ii) Change to fuels which give low or zero amounts of CO_2 to the atmosphere. All the renewable sources of energy (wind, waves, solar, hydro) would seem to be essentially CO_2-free, as is nuclear power. As is evident from Table 5, some fossil fuels produce more CO_2 per unit of energy than others. Thus by altering the mix to fossil fuels used (more gas and oil, less coal) CO_2 emission savings could be made. However, the potential here is limited in the longer term because of the relatively small reserves of gas and oil compared with coal.

(iii) Planting trees. It has been shown in Chapter 3 that, unless accompanied by replanting, cutting of forests results in a net flux of CO_2 into the atmosphere since the carbon stored in tree wood and forest soil is subject to oxidation after clearance. In contrast, by planting new forests the amount of carbon stored in the biosphere will increase at the expense of CO_2 in the atmosphere. However, the amounts of new planting to be achieved are such as to mean that afforestation could never be more than a very partial answer to the problem. At present rates of fossil fuel combustion (5 GtC $year^{-1}$, up to half of which enters the oceans), in order to back off the CO_2 build-up in the atmosphere at least 2.5 GtC would need to be sequestered in additional forest (or other long-term) growth each year. At first sight it might seem that a total living biomass estimated to contain between 450 and 1000 GtC (see Ajtay et al., 1979) could easily take up this relatively small additional increment. Such a comparison is, however, not valid since the annual long-term storage of carbon by the biosphere (balanced in the steady state by oxidation and decay of dead organic matter) is only a small fraction of the biomass. Of the carbon absorbed in photosynthesis (the Gross Primary Production (GPP)) about half is returned to the atmosphere through respiration. Of the remainder (the Net Primary Production (NPP)) a considerable portion, depending on the stage of growth, will be used in the production of annual litter which is quickly returned to the atmosphere via the animal food chain or microbial decay. The residue is available for long-term storage as wood or as soil organic matter and is termed the Net Ecosystem Production (NEP). To quantify the picture we note that Woodwell et al. (1978) estimate that in a total living biomass of 826 GtC, 743 GtC are contained in the world's forests, indicating that trees must be responsible for a large part of the NEP. They further calculate the NPP of forests to be 33 GtC year^{-1}. Hampicke (1979) quotes a NEP/NPP ratio of 0.4 for temperate forests, which, if applied to all forests, suggests that only 13 GtC are stored annually. If it were planned to increase the NEP by 2.5 GtC year^{-1} an increase in the forested area approaching 20% would be required. Approaching the calculation from a different angle, we make use of the fixation rate of 7.5 x 10^8 gC km^{-2} year^{-1} estimated by Steinbeck and Brown (1976) for American Sycamore grown on marginal land. To account for 2.5 GtC year^{-1}, an area of 3 x 10^6 km^2 of new forest would need to be planted. This is roughly the size of India and represents 6 or 10% of the present forest area, depending on whether the latter is taken as 31 x 10^6 km^2 (Ajtay et al., 1979) or 49 x 10^6 km^2 (Woodwell et al., 1978). One reason why this estimate of the required expansion is less than that calculated above may be that an optimum assimilation rate, corresponding with mid-term growth, has been used, whereas the NEP/NPP ratio used earlier is more representative of existing forests. Whichever estimate is nearer to the true requirement, however, it seems most unlikely that such an extensive planting programme could ever be mounted. Even if areas suitable for aforestation were made available, it is probable that there would be considerable pressures in favour of increased food production when decisions on land-use came to be made. Furthermore, the above calculations make no allowance for future increases in fossil fuel usage.

(iv) Remove CO_2 from emissions coming from fixed sources. By analogy with the removal of SO_2 from power station flue gases, so it has been suggested that it might be technically feasible to remove CO_2 from such sources. If this is to be done then not only does the CO_2 have to be removed from the flue gas but also it has to be disposed of in such a way that the CO_2 is not returned to the atmosphere. For example, there is no point in scrubbing CO_2 out of the flue gases into a solvent (water) if the water is then discharged to a surface environment, where it finds itself supersaturated with respect to the CO_2 partial pressure in the atmosphere, since the CO_2 will simply degas back into the air. To deal with this problem Marchetti (1977) proposed injecting the collected CO_2 into seawater at places where deep ocean water is being formed naturally, and suggested the Straits of Gibraltar, where Mediterranean water sinks to reach a depth of about 1 km in the Atlantic, as a possible location. Others (for example Goldthorpe and Jorro, 1979; Albanese and Steinberg, 1980) investigate the use of seawater itself as the scrubbing fluid which is then injected into the subsurface oceans. These authors argue that the costs both in terms of energy (on some calculations the CO_2 extraction plant requires more energy than is produced by the power station generating the gas which is to be scrubbed) and money are very unfavourable. Somewhat less unfavourable is the option of using monoethanolamine (MEA) to extract the CO_2 from the flue gases according to the following reaction,

$$HO(CH_2)_2\ NH_2 + H_2O + CO_2 \rightarrow HO(CH_2)_2\ NH_3HCO_3 \quad ...(19)$$

The MEA is then regenerated and the CO_2 it has absorbed compressed and transported to the coast for dumping into the deep oceans. Baes et al. (1980) have calculated that use of the MEA process to remove 90% of the CO_2 together with compressing and transporting it to a coastal site assumed to be 40 km from the power plant would impose an energy penalty of near 50% of the combustion energy of the fuel used in the power station. The cost of the operation is given as \$24 per 10^6 g of CO_2 removed, which corresponds to about a 50% money penalty at present generating costs in the U.K. equivalent to about 4cents kWh^{-1}. The energy and financial penalties combine to increase the final cost of the electricity transmitted by a factor of three. These estimates do not appear to include the costs of getting the CO_2 into the deep oceans. An interesting idea attributed to Marchetti and discussed by Baes et al. (1980) is to combust fossil fuels in pure oxygen, rather than air, to produce a flue gas consisting almost entirely of CO_2 and H_2O. The cost of extracting O_2 from air is outweighed by the advantage of not having to scrub CO_2 from the flue gas since it can be compressed directly. Another potential benefit of such a scheme would be that SO_2 and any oxides of nitrogen (which latter would be much reduced in any case) could probably be recovered along with the CO_2.

It should be noted that all the options discussed refer only to removal of CO_2 from large fixed sources, such as power stations. Even in these favourable circumstances very substantial energy and money costs are incurred. It seems extremely unlikely that the removal option would ever be applicable to smaller scale sources which constitute two thirds of total CO_2 emissions.

Although the various options discussed here, taken singly, are either costly, impracticable, or can solve only a small fraction of the problem, some combination of them may constitute a valid response both in terms of magnitude and cost. Furthermore, even without any link between CO_2 emissions and climate several of the options discussed are being put into practice, or may well be in the future, for very different reasons. For example, increased fossil fuel costs have certainly led to more efficient use, alternative (and largely non-CO_2 producing) sources of power are either being operated and actively installed (nuclear) or intensively researched (solar, wind, tidal etc.), and in some parts of the world the trend is towards net planting rather than cutting of trees.

In the CO_2 context such developments are to be encouraged since they will make a positive contribution to the solution of any CO_2-induced climate problem which is recognised in the future.

Global CO_2 research effort

The proportions of global CO_2 production from fossil fuel burning attributable to 12 geographical areas of the world have been presented by Marland and Rotty (1980) for the year 1974 (Fig. 37(a)). Together the USA, Europe and the USSR account for almost three quarters of the world production. Of the 18% contributed by Western Europe, the UK is responsible for about one fifth.

Using estimates of projected populations, size of labour force, productivity trends and energy to economy relations, Rotty and Marland (1980) calculated energy demands for various regions of the world. From the calculations the partitioning of CO_2 emissions for the year 2025 was predicted and was displayed in the form shown in Fig. 37(b) by Marland and Rotty (1980). Based on this particular projection (one of the many and very varied predictions to have appeared in the literature) it is seen that the contribution from the Western World becomes much less significant by the early decades of the next century. However, Kellogg and Schware (1981) point out that if *cumulative* emissions are considered, the integrated contribution from the developed nations is likely to remain well above that from the developing world (Developing + CTP Asia in Fig. 37(b)) for most of the next century. Thus despite a continual decrease in the proportion of emissions, the developed world will continue to carry the major responsibility for the accumulation of CO_2. It is clear, also, that if ameliorative measures were ever considered necessary, they would only be effective if pursued by many nations acting together.

As highlighted throughout this book, much research needs to be done in order to reach more definitive answers to the many unresolved questions involved in the CO_2/climate issue. In the United States, the National Climate Program Office launched, in the late 1970s, a National Program on Carbon Dioxide, Environment and Society. In 1981 (L. Machta, personal communication) the total funding of \$19.5M was divided into \$12.1M for carbon cycle research, \$3.6M for CO_2/climate research, \$1.9M for research into vegetation effects, \$0.4M for the assessment of socio-economic effects and \$0.6M for studying impact amelioration, with the balance on integration, management and evaluation. About half the funding comes from the US Department of Energy, who coordinate the activities of the programme. In some other countries (e.g. Canada, Australia) attempts have been made to coordinate the national CO_2 research effort, but in others, including the UK, it is difficult to ascertain what is being done and whether or not the funding is sufficient, since no 'machinery' exists for monitoring the diverse range of interdisciplinary research

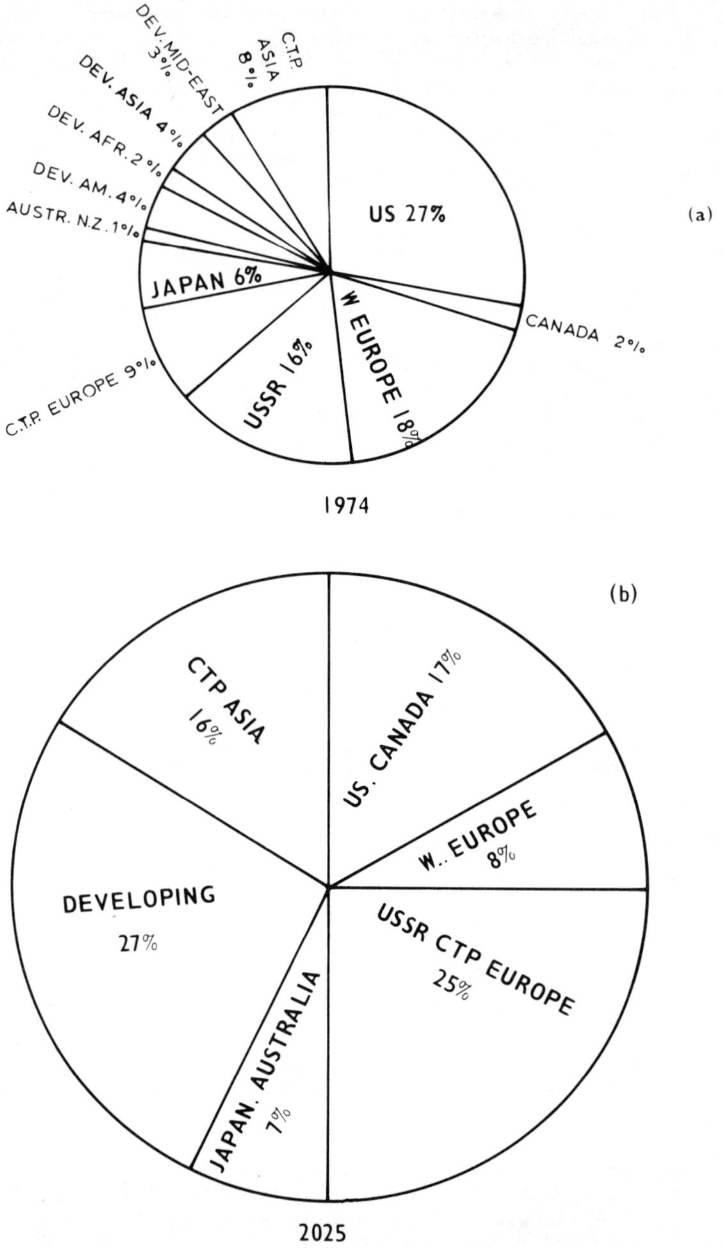

Figure 37. Global CO_2 production by world segments (a) for 1974 and (b) estemated for 2025. (CTP ≡ centrally planned economies). Disc areas in proportion to total production. (After Marland and Rotty, 1980).

relevant to the problem. Inactivity in this latter regard perhaps indicates a less than adequate interest.

Since the CO_2 issue is of global concern research must be conducted in an international context. Recognizing this, the World Climate Programme, organised jointly by WMO, UNEP and ICSU, has identified the problem as one of its major interests. However the efficiency with which it can tackle the problem must depend to a large extent on the collaboration of the relevant national governmental bodies and industries with the World Climate Programme. In turn such collaboration depends on the ability of individual nations to coordinate their own internal efforts, such as is being attempted in the USA. This should entail not only monitoring current intra-national CO_2 research, but also being aware of relevant expertise in disciplines impinging on the CO_2 issue, which exists in their own countries, but which is not currently being tapped. Only by being so prepared will they be able to play an adequate role in international cooperation.

PART IV

Summary

In the unperturbed environment CO_2 is cycled across the air-sea and land-air boundaries in substantial amounts, with transfers across these interfaces in approximate balance on a yearly average basis. By combustion of fossil fuels, and probably also by land-use changes, man has mobilized significant additional amounts of CO_2, the majority of which has been injected directly into the atmosphere. Measurements made over the last 25 years at the Mauna Loa Observatory, Hawaii, and over shorter periods elsewhere, show clearly an increase in the atmospheric concentration of CO_2 of about 0.3% $year^{-1}$, which is attributed to these anthropogenic inputs. Quantitatively, over the period 1958 to the present, for which accurate records are available, the observed CO_2 increase corresponds to only about half of the amount emitted from fossil fuel combustion. Model calculations indicate that uptake of CO_2 by the oceans can probably account for most of the remainder, but if the biosphere has been a net source then either the ocean sink must be larger than previously thought or other sinks, as yet poorly identified, must be significant.

Due to this uncertainty and those associated with forecasting trends in fossil fuel use, prediction of atmospheric CO_2 levels in the future is uncertain. Best estimates, based on a 2% $year^{-1}$ increase in fossil fuel usage, indicate that a doubling of the pre-industrial atmospheric concentration will probably occur sometime between 2050 and 2075, although growth in fossil fuel use at rates which occurred in the period 1860-1970 (approximately 4% $year^{-1}$) would lead to doubling as early as 2035.

The presence of naturally occurring CO_2 in the atmosphere leads to an elevation in the Earth's mean surface temperature of order 10°C through the 'greehouse effect'. It is currently estimated, through the use of global climate models, that a doubling in atmospheric CO_2 by man's activities would lead to an increase in mean surface temperature of 2-3°C, with warming of up to 6°C in high latitudes. Such changes would be greater than any experienced naturally over the last 10 000 years. Associated with these temperature changes would likely be significant alterations in rainfall distribution.

Considerable uncertainty exists in these estimates, due mainly to inadequate modelling of changes in cloudiness and in the way the oceans take up and redistribute heat. A number of plausible feedback processes involving cloud characteristics which are not modelled at present have been suggested as mechanisms by which the warming would be reduced. However, it is equally possible to postulate other feedback processes which could enhance the warming. It would therefore be extremely unwise to assume that the present estimates of climatic change are excessive and that future developments in modelling will necessarily lead to predictions of less drastic warming.

Although detailed regional variations of a CO_2-induced climatic change cannot yet be simulated, it seems probable that some areas of the world would benefit from the climatic changes while for other regions they would be detrimental. In either case, however, the climatic transition would be disruptive of agricultural and other climate-related systems, since these tend to be tuned to present-day climates.

Although the potential problem of CO_2-induced climatic change has been known for many years, it is only in the last decade that serious research efforts have been made for its elucidation. This has come about essentially because of the existence of accurate records of atmospheric CO_2 increase and, in view of the modelling uncertainties, it can be argued that until an analogous signal is observable in the climatic record, which will probably not occur until about the turn of the century, it is unlikely that the issue of whether or not to counteract the CO_2 increase will be seriously considered. Assuming the climate model predictions are accurate, this state of affairs is unsatisfactory since, if a decision to change from fossil fuel to other energy technologies were only then deemed necessary, a substantial CO_2-induced climatic warming would have already occurred by the time the change in fuel use became fully effective in the mid-21st century. In view of this timescale problem, it is clearly necessary for research to be pursued actively in order to narrow present uncertainties, thereby improving the reliability of the predictions of climatic change and enabling earlier decisions to be made regarding possible counteractive measures. Such research needs to concentrate on the time dependence of CO_2-induced climatic changes which will require, in addition to improved modelling capability, a better prediction of future atmospheric CO_2 levels. In turn, the latter depends on a better knowledge of the partitioning of anthropogenically released CO_2 between the atmosphere, the oceans and other natural sinks.

If it should be judged necessary to control the increase of atmospheric CO_2 then several potential options are available. The least feasible of those proposed appears to be direct removal of CO_2 from industrial combustion processes and increased fixation of carbon by afforestation. The former suffers from substantial penalties in terms of energy used in extracting the CO_2 and the cost thereof, as well as possibly insurmountable difficulties with regard to the ultimate disposal of the collected carbon dioxide. Afforestation to the extent required even to remove present anthropogenic CO_2 emissions seems to require more land than is conceivably available. This leaves either a reduction in energy use (assuming a fuel mix as at present) or change to non-fossil fuel energy sources, or a combination of both, as the only practical options.

It is evident that the continuing increase in global CO_2 emissions is a potentially serious problem which could significantly affect industrial operations. The CO_2 problem is by its nature a global concern and research must be conducted in an international context. It is therefore arguable that relevant international organisations (e.g. UNEP, WMO, ICSU) should be encouraged to seek greater collaboration of government bodies and energy producers in order to rationalize and increase the total research effort on the problem, thus reducing uncertainties and allowing proper economic energy planning to proceed. At the same time, the pursuance of economically sound energy efficiency and conservation programmes and the development of alternative energy technologies, both of which are needed even in the absence of a CO_2 problem, should be encouraged as parallel measures.

References

Adams, J.A.S., Mantovani, M.S.M. and Lundell, L.L., 1977. Wood versus fossil fuel as a source of excess carbon dioxide in the atmosphere: a preliminary report. *Science*, 196, 54-56.

Ajtay, G.L., Ketner, P. and Duvigneaud, P., 1979. Terrestrial primary production and phytomass, In: *SCOPE 13 - The Global Carbon Cycle*, Ed. B. Bolin, E.T. Degens, S. Kempe and P. Ketner, J. Wiley and Sons, 129-181.

Albanese, A.S. and Steinberg, M., 1980. Environmental control technology for atmospheric carbon dioxide. *Energy*, 5, 641-664.

Augustsson, T. and Ramanathan, V., 1977. A radiative-convective model study of the CO_2 climate problem. *Journal of the Atmospheric Sciences*, 34, 448-451.

Bacastow, R.B. and Bjorkstrom, A., 1981. Comparison of ocean models for the carbon cycle, In: *SCOPE 16 - Carbon Cycle Modelling*, Ed. B. Bolin, J. Wiley and Sons, 29-80.

Bacastow, R.B. and Keeling, C.D., 1981. Atmospheric carbon dioxide concentration and the observed airborne fraction, In: *SCOPE 16 - Carbon Cycle Modelling*, Ed. B. Bolin, J. Wiley and Sons, 103-112.

Baes, C.F., Beall, S.E. and Lee, D.W., 1980. The collection, disposal and storage of carbon dioxide, In: *Interactions of Energy and Climate*, Ed. W. Bach, J. Pankrath and J. Williams, D. Reidel Pub. Co., 495-519.

Bakan, S., 1978. Note on the eddy correlation method for CO_2 flux measurements. *Boundary Layer Meteorology*, 14, 597-600.

Balzer, W. and Wefer, G., 1981. Dissolution of carbonate minerals in a subtropical shallow marine environment. *Marine Chemistry*, 10, 545-558.

Barrett, E.W., 1975. Inadvertant weather and climate modification. *Critical Reviews in Environmental Control*, 6, 15-90.

Berner, W., Oeschger, H. and Stauffer, B., 1980. Information on the CO_2 cycle from ice core studies. *Radiocarbon*, 22, 227-235.

Bischof, W., 1960. Periodical variations of the atmospheric CO_2-content in Scandinavia. *Tellus*, 12, 216-226.

Bjorkstrom, A., 1979. A model of CO_2 interaction between atmosphere, oceans and land biota. In: *SCOPE 13 - The Global Carbon Cycle*, Ed. B. Bolin, E.T. Degens, S. Kempe and P. Ketner, J. Wiley and Sons, 403-457.

Bolin, B., 1977. Changes of land biota and their importance for the carbon cycle. *Science*, 196, 613-615.

Bolin, B., Degens, E.T., Duvigneaud, P. and Kempe, S., 1979. The global biogeochemical carbon cycle. In: *SCOPE 13 - The Global Carbon Cycle*, Ed. B. Bolin, E.T. Degens, S. Kempe and P. Ketner, J. Wiley and Sons, 1-56.

Bray, J.R., 1959. An analysis of the possible recent change in atmospheric carbon dioxide concentration. *Tellus*, 11, 220-230.

Brewer, P.G., 1978. Direct observation of the oceanic CO_2 increase. *Geophysical Research Letters*, 5, 997-1000.

Broecker, W.S., 1980a. Status of $^{13}C/^{12}C$ research on tree rings. In: *Proceedings of the International Meeting on Stable Isotopes in Tree-ring Research*, Ed. G. Jacoby, US Dept. of Energy, 6-9.

Broecker, W.S., 1980b. Glacial to interglacial changes in ocean chemistry. In: *CIMAS Symposium Volume*, Ed. E. Kraus, University of Miami.

Broecker, W.S., Peng, T.-H. and Engh, R., 1980a. Modelling the carbon system. *Radiocarbon*, 22, 565-598.

Broecker, W.S., Peng, T.-H. and Stuiver, M., 1978. An estimate of the upwelling rate in the equatorial atlantic based on the distribution of bomb radiocarbon. *Journal of Geophysical Research*, 83, 6179-6186.

Broecker, W.S., Peng, T.-H. and Takahashi, T., 1980b. A strategy for the use of bomb-produced radiocarbon as a tracer for the transport of fossil fuel CO_2 into the deep-sea source regions. *Earth and Planetary Science Letters*, 49, 463-468.

Broecker, W.S., Takahashi, T., Simpson, H.J. and Peng, T.-H., 1979. Fate of fossil fuel carbon dioxide and the global carbon budget. *Science*, 206, 409-418.

Bryan, K., Komro, F.G., Manabe, S. and Spelman, M.J., 1982. Transient climate response to increasing atmospheric carbon dioxide. *Science*, 215, 56-58.

Callendar, G.S., 1938. The artificial production of carbon dioxide and its influence on temperature. *Quarterly Journal of the Royal Meteorological Society*, 64, 223-240.

Callendar, G.S., 1958. On the amount of carbon dioxide in the atmosphere. *Tellus*, 10, 243-248.

Cess, R.D. and Goldenberg, S.D., 1981. The effect of ocean heat capacity upon global warming due to increasing carbon dioxide. *Journal of Geophysical Research*, 86, 498-502.

Chen, G.-T. and Millero, F.J., 1979. Gradual increase of oceanic CO_2. *Nature*, 277, 205-206.

Coakley, J.A. and Wielicki, B.A., 1979. Testing energy balance climate models. *Journal of the Atmospheric Sciences*, 36, 2031-2039.

Craig, H., 1957. The natural distribution of radiocarbon and the exchange time of carbon dioxide between the atmosphere and the sea. *Tellus*, 9, 1-17.

Crane, A.J., 1981. Comments on recent doubts about the CO_2 greenhouse effect. *Journal of Applied Meteorology*, 20, 1547-1549.

Crane, A.J., 1982. The partitioning of excess CO_2 in a five-reservoir atmosphere-ocean model. *Tellus*, 34, 398-405.

Delcourt, H.R. and Harris, W.F., 1980. Carbon budget of the southeastern US biota: analysis of historical changes in trend from source to sink. *Science*, 210, 321-323.

Delmas, R.J., Ascencio, J.-M. and Legrand, M., 1980. Polar ice evidence that atmospheric CO_2 20 000 years BP was 50% of present. *Nature*, 284, 155-157.

Downton, W.J.S., Bjorkman, O. and Pike, C.S., 1980. Consequences of increased atmospheric concentrations of carbon dioxide for growth and photosynthesis of higher plants. In: *Carbon Dioxide and Climate: Australian Research*, Ed. G.I. Pearman, Australian Academy of Science, 143-151.

Fisher, B.E.A. and Macqueen, J.F., 1980. The influence of the oceans on the atmospheric burden of carbon dioxide. *Applied Mathematical Modelling*, 4, 439-448.

Francey, R.J., 1981. Tasmanian tree rings belie suggested anthropogenic $^{13}C/^{12}C$ trends. *Nature*, 290, 232-235.

Freyer, H.D., 1978. Preliminary evaluation of past CO_2 increase as derived from ^{13}C measurements in tree rings. In: *Carbon Dioxide, Climate and Society*, Ed. J. Williams, Pergamon Press, 69-77.

Freyer, H.D., 1979. Variations in the atmospheric CO_2 content. In: *SCOPE 13 - The Global Carbon Cycle*, Ed. B. Bolin, E.T. Degens, S. Kempe and P. Ketner, J. Wiley and Sons, 79-99.

Gates, W.L., Cook, K.H. and Schlesinger, M.E., 1981. Preliminary analysis of experiments on the climatic effects of increased CO_2 with an atmospheric general circulation model and a climatological ocean. *Journal of Geophysical Research*, 86, 6385-6393.

Goldthorpe, S.H. and Jorro, M.A.A., 1979. *Could CO_2 be Removed at Source? - An Assessment*. Internal Report, National Coal Board.

Goudriaan, J. and Ajtay, G.L., 1979. The possible effects of increased CO_2 on photosynthesis. In: *SCOPE 13 - The Global Carbon Cycle*, Ed. B. Bolin, E.T. Degens, S. Kempe, P. Ketner, J. Wiley and Sons, 237-249.

Hall, C.A.S., Ekdahl, C.A. and Wartenberg, D.E., 1975. A fifteen-year record of biotic metabolism in the Northern Hemisphere. *Nature*, 255, 136-138.

Hampicke, V., 1979. Net transfer of carbon between the land biota and the atmosphere induced by man. In: *SCOPE 13 - The Global Carbon Cycle*, Ed. B. Bolin, E.T. Degens, S. Kempe and P. Ketner, J. Wiley and Sons, 219-236.

Hanel, R.A., Schlachman, B., Rodgers, D. and Vanous, D., 1971. Nimbus 4 Michelson interferometer. *Applied Optics*, 10, 1376-1382.

Hansen, J., Johnson, D., Lacis, A., Lebedeff, S., Lee, P., Rind, D. and Russell, G., 1981. Climate impact of increasing atmospheric carbon dioxide. *Science*, 213, 957-966.

Hansson, I., 1973. A new set of acidity constants for carbonic acid and boric acid in sea water. *Deep-Sea Research*, 29, 461-478.

Hartmann, D.L. and Short, D.A., 1980. On the use of earth radiation budget statistics for studies of clouds and climate. *Journal of the Atmospheric Sciences*, 37, 1233-1250.

Hasse, L. and Liss, P.S., 1980. Gas exchange across the air-sea interface. *Tellus*, 32, 470-481.

Hoffert, M.I., Callegari, A.J. and Hsieh, C.-T., 1980. The role of deep sea heat storage in the secular response to climatic forcing. *Journal of Geophysical Research*, 85, 6667-6679.

Houghton, J.T., 1977. *The Physics of Atmospheres*. Cambridge Univ. Press, 203 pp.

Hunt, B.G., 1981. An examination of some feedback mechanisms in the carbon dioxide climate problem. *Tellus*, 33, 78-88.

Hunt, B.G. and Wells, N.C., 1979. An assessment of the possible future impact of carbon dioxide increases based on a coupled one-dimensional atmospheric-oceanic model. *Journal of Geophysical Research*, 84, 787-791.

Idso, S.B., 1980. The climatological significance of a doubling of the earth's atmospheric carbon dioxide concentration. *Science*, 207, 1462-1463.

Jenkins, W.J., 1980. Tritium and ^{3}He in the Sargaso Sea. *Journal of Marine Research*, 38, 533-569.

Jones, E.P. and Smith, S.D., 1977. A first measurement of sea-air CO_2 flux by eddy correlation. *Journal of Geophysical Research*, 82, 5990-5992.

Keeling, C.D., 1973(a). Industrial production of carbon dioxide from fossil fuels and limestone. *Tellus*, 25, 174-198.

Keeling, C.D., 1973(b). The carbon dioxide cycle. In: *Chemistry of the Lower Atmosphere*, Ed. S.I. Rasool, Plenum Press, 251-331.

Kellogg, W.W., 1977. *Effects of Human Activities on Global Climate*. WMO Tech Note No. 156, WMO No. 486, Geneva.

Kellogg, W.W., 1978. Global influences of mankind on the climate. In: *Climatic Change*, Ed. J. Gribbin, Cambridge Univ. Press, 205-227.

Kellogg, W.W., 1980. Aerosols and climate. In: *Interactions of Energy and Climate*, Ed. W. Bach, J. Pankrath and J. Williams, D. Reidel Pub. Co., 281-296.

Kellogg, W.W. and Schware, R., 1981. *Climate Change and Society*. Westview Press, Boulder, Colorado, 178 pp.

Knutzen, J., 1981. Effects of decreased pH on marine organisms. *Marine Pollution Bulletin*, 12, 25-29.

Kramer, P.J., 1981. Carbon dioxide concentration, photosynthesis, and dry matter production. *Bioscience*, 31, 29-33.

Lacis, A., Hansen, J., Lee, P., Mitchell, T. and Lebedeff, S., 1981. Greenhouse effect of trace gases, 1970-1980. *Geophysical Research Letters*, 8, 1035-1038.

Lamb, H.H., 1970. Volcanic dust in the atmosphere. *Philosophical Transactions of the Royal Society of London*, A 266, 426-453.

Lemon, E., 1977. The land's response to more carbon dioxide. In: *The Fate of Fossil Fuel* CO_2 *in the Oceans*, Ed. N.R. Andersen and A. Malahoff, Plenum Press, 97-130.

Lian, M.S., 1978. *Energy Balance Climate Models and Climate Sensitivity*, Ph.D. Thesis, State University of New York, Stony Brook.

Liss, P.S., 1983. Gas transfer: experiments and geochemical implications. In: *Air-Sea Exchange of Gases and Particles*, Ed. P.S. Liss and W.G.N. Slinn, D. Reidel Pub. Co., in press.

Loomis, R.S., 1979. CO_2 and the biosphere. In: *Workshop on the Global Effects of Carbon Dioxide from Fossil Fuels*, Ed. W.P. Elliott and L. Machta, US Dept. of Energy, 51-62.

Loucks, O.L., 1980. Recent results from studies of carbon cycling in the biosphere. In: *Proceedings of the Carbon Dioxide and Climate Research Program Conference*, Ed. L.E. Schmitt, US Dept. of Energy, 3-42.

Lovins, A.B., Lovins, A.H., Krause, F. and Bach, W., 1982. *Least-cost Energy : Solving the* CO_2 *Problem*. Brickhouse Publishing.

Machta, L., 1972. The role of the oceans and biosphere in the carbon dioxide cycle. In: *The Changing Chemistry of the Oceans*, Ed. D. Dyrssen and D. Jagner, J. Wiley and Sons, 121-145.

Machta, L., 1979. Atmospheric measurements of carbon dioxide. In: *Workshop on the Global Effects of Carbon Dioxide from Fossil Fuels*, Ed. W.P. Elliott and L. Machta, US Dept. of Energy, 44-50.

Mackenzie, F.T., 1981. Global carbon cycle: some minor sinks for CO_2. In: *Workshop on the Flux of Organic Carbon by Rivers to the Oceans*, US Dept. of Energy, 360-384.

Madden, R.A. and Ramanathan, V., 1980. Detecting climate change due to increasing carbon dioxide. *Science*, 209, 763-768.

Manabe, S., Bryan, K. and Spelman, M.J., 1979. A global ocean-atmosphere climate model with seasonal variation for future studies of climate sensitivity. *Dynamics of Atmospheres and Oceans*, 3, 393-426.

Manabe, S. and Stouffer, R.J., 1979. A CO_2-climate sensitivity study with a mathematical model of the global climate. *Nature*, 282, 491-493.

Manabe, S. and Stouffer, R.J., 1980. Sensitivity of a global climate model to an increase of CO_2 concentration in the atmosphere. *Journal of Geophysical Research*, 85, 5529-5554.

Manabe, S. and Strickler, R.F., 1964. Thermal equilibrium of the atmosphere with a convective adjustment. *Journal of the Atmospheric Sciences*, 21, 361-385.

Manabe, S. and Wetherald, R.T., 1967. Thermal equilibrium of the atmosphere with a given distribution of relative humidity. *Journal of the Atmospheric Sciences*, 24, 241-259.

Manabe, S. and Wetherald, R.T., 1975. The effect of doubling the CO_2 concentration on the climate of a general circulation model. *Journal of the Atmospheric Sciences*, 32, 3-15.

Manabe, S. and Wetherald, R.T., 1980. On the distribution of climate change resulting from an increase in CO_2 content of the atmosphere. *Journal of the Atmospheric Sciences*, 37, 99-118.

Manabe, S., Wetherald, R.T. and Stouffer, R.J., 1981. Summer dryness due to an increase of atmospheric CO_2 concentration. *Climatic Change*, 3, 347-385.

Marchetti, C., 1977. On geoengineering and the CO_2 problem. *Climatic Change*, 1, 59-68.

Marland, G. and Rotty, R.M., 1980. Atmospheric carbon dioxide: What to do? *Consensus*, 2-3, 41-52.

McKay, G.A. and Allsopp, T., 1980. The role of climate in affecting energy demand/supply. In: *Interactions of Energy and Climate*, Ed. W. Bach, J. Pankrath and J. Williams, D. Reidel Pub. Co., 53-72.

Mitchell, J.F.B., 1983. The seasonal response of a general circulation model to changes in CO_2 and sea temperatures. *Quarterly Journal of the Royal Meteorological Society*, 109, 113-152.

Monteith, J.L., 1981. Climatic variation and the growth of crops. *Quarterly Journal of the Royal Meteorological Society,* 107, 749-774.

Moore, B., Boone, R.D., Hobbie, J.E., Houghton, R.A., Melillo, J.M., Peterson, B.J., Shaver, G.R., Vorosmarty, C.J. and Woodwell, G.M., 1981. A simple model for analysis of the role of terrestrial ecosystems in the global carbon budget. In: *SCOPE 16 - Carbon Cycle Modelling,* Ed. B. Bolin, J. Wiley and Sons, 365-385.

National Academy of Sciences, 1979. *Carbon Dioxide and Climate: A Scientific Assessment*. Report of an Ad Hoc NRC study group on CO_2 and climate chaired by J.G. Charney, Woods Hole, Mass., 23-27 July 1979, National Academy Press, Washington, DC.

National Academy of Sciences, 1982. *Carbon Dioxide and Climate: A Second Assessment*. Report of an Ad Hoc NRC study group on CO_2 and climate chaired by J. Smagorinsky, National Academy Press, Washington, DC.

Neftel, A. Oeschger, H., Schwander, J., Stauffer, B. and Zumbrunn, R., 1982. Ice core sample measurements give atmospheric CO_2 content during the past 40 000 years. *Nature,* 295, 220-223.

Newell, R.E. and Dopplick, T.G., 1979. Questions concerning the possible influence of anthropogenic CO_2 on atmospheric temperature. *Journal of Applied Meteorology,* 18, 822-825.

Oeschger, H. and Heimann, M., 1981. Uncertainties of predictions of future atmospheric CO_2 concentrations, Presented at *WMO/ICSU/UNEP Meeting on Analysis and Interpretation of Atmospheric CO_2 Data, Berne, 14-18/9/81,* WCP-14, Geneva.

Oeschger, H., Siegenthaler, U., Schotterer, U. and Gugelmann A., 1975. A box diffusion model to study the carbon dioxide exchange in nature. *Tellus,* 27, 168-192.

Ohring, G. and Clapp, P., 1980. The effect of changes in cloud amount on the net radiation at the top of the atmosphere. *Journal of the Atmospheric Sciences,* 37, 447-454.

Paltridge, G.W., 1980. Cloud-radiation feedback to climate. *Quarterly Journal of the Royal Meteorological Society,* 106, 895-899.

Paltridge, G.W. and Platt, C.M.R., 1976. *Radiative Processes in Meteorology and Climatology*. Developments in Atmospheric Science, 5, Elsevier Scientific Publishing Co.

Park, P.K., 1969. Oceanic CO_2 system. An evaluation of ten methods of investigation. *Limnology and Oceanography,* 14, 179-186.

Pearman, G.I. and Hyson, P. 1981. The annual variation of atmospheric CO_2 concentration observed in the Northern Hemisphere. *Journal of Geophysical Research,* 86, 9839-9843.

Peng, T.-H., Broecker, W.S., Mathieu, G.G., Li, Y.H. and Bainbridge, A.E., 1979. Radon evasion rates in the Atlantic and Pacific Oceans as determined during the GEOSECS program. *Journal of Geophysical Research*, 84, 2471-2486.

Ralston, C.W., 1979. Where has all the carbon gone? *Science*, 204, 1345-1346.

Ramanathan, V., 1980. Climatic effects of anthropogenic traces gases. In: *Interactions of Energy and Climate*, Ed. W. Bach, J. Pankrath and J. Williams, D. Reidel Pub. Co., 269-280.

Ramanathan, V., 1981. The role of ocean-atmosphere interactions in the CO_2 climate problem. *Journal of the Atmospheric Sciences*, 38, 918-930.

Ramanathan, V. and Coakley, J.A., 1978. Climate modelling through radiative-convective models. *Reviews of Geophysics and Space Physics*, 16, 465-489.

Ramanathan, V., Lian, M.S. and Cess, R.D., 1979. Increased atmospheric CO_2: zonal and seasonal estimates of the effect on the radiation energy balance and surface temperature. *Journal of Geophysical Research*, 84, 4949-4958.

Roach, W.T. and Slingo, A., 1979. A high resolution infrared radiative transfer scheme to study the interaction of radiation with cloud. *Quarterly Journal of the Royal Meteorological Society*, 105, 603-614.

Rooth, C.G. and Ostlund, H.G., 1972. Penetration of tritium into the Atlantic Thermocline. *Deep-Sea Research*, 19, 481-492.

Rosenberg, N.J., 1981. The increasing CO_2 concentration in the atmosphere and its implication on agricultural productivity. *Climatic Change*, 3, 265-279.

Rotty, R.M., 1973. Commentary on and extension of calculative procedures for CO_2 production. *Tellus*, 25, 508-517.

Rotty, R.M., 1974. First estimates of global flaring of natural gas. *Atmospheric Environment*, 8, 681-686.

Rotty, R.M., 1977. Global carbon dioxide production from fossil fuels and cement, A.D. 1950 - A.D. 2000. In: *The Fate of Fossil Fuel CO_2 in the Oceans*, Ed. N.R. Andersen and A. Malahoff, Plenum Press, 167-181.

Rotty, R.M., 1979. Present and future production of CO_2 from fossil fuels - a global appraisal. In: *Workshop on the Global Effects of Carbon Dioxide from Fossil Fuels*, Ed. W.P. Elliott and L. Machta, US Dept. of Energy, 36-43.

Rotty, R.M., 1980. Uncertainties associated with global effects of atmospheric CO_2. *Science of the Total Environment*, 15, 73-86.

Rotty, R.M., 1983. Distribution of and changes in industrial carbon dioxide production. *Journal of Geophysical Research,* 88, 1301-1308.

Rotty, R.M. and Marland, G., 1980. Constraints on fossil fuel use. In: *Interaction of Energy and Climate,* Ed. W. Bach, J. Pankrath and J. Williams, D. Reidel Pub. Co., 191-212.

Rowntree, P.R. and Walker, J., 1978. The effects of doubling the CO_2 concentration on radiative-convective equilibrium. In: *Carbon Dioxide Climate and Society,* Ed. J. Williams, Pergamon Press, 181-191.

Schneider, S.H., 1975. On the carbon dioxide - climate confusion. *Journal of the Atmospheric Sciences,* 32, 2060-2066.

Schneider, S. and Thompson, S.L., 1981. Atmospheric CO_2 and climate: Importance of the transient response. *Journal of Geophysical Research,* 86, 3135-3147.

Schneider, S.M., Washington, W.M. and Chervin, R.M., 1978. Cloudiness as a climate feedback mechanism: effects on cloud amounts of prescribed global and regional surface temperature changes in the NCAR GCM. *Journal of the Atmospheric Sciences,* 35, 2207-2221.

Seiler, W. and Crutzen, P.J., 1980. Estimates of gross and net fluxes of carbon between the biosphere and the atmosphere from biomass burning. *Climatic Change,* 2, 207-247.

Sellers, W.D., 1974. A reassessment of the effect of CO_2 variations on a simple global climate model. *Journal of Applied Meteorology,* 13, 831-833.

Siegenthaler, U., 1983. Uptake of excess CO_2 by an outcrop diffusion model of the ocean. *Journal of Geophysical Research,* 88, 3599-3608.

Siegenthaler, U., Heimann, M. and Oeschger, H., 1978. Model response of the atmospheric CO_2 level and $^{13}C/^{12}C$ ratio to biogenic CO_2 input. In: *Carbon Dioxide, Climate and Society,* Ed. J. Williams, Pergamon Press, 79-87.

Siegenthaler, U. and Oeschger, H., 1978. Predicting future atmospheric carbon dioxide levels. *Science,* 199, 388-395.

Steinbeck, K. and Brown, C.L., 1976. Yield and utilization of hardwood fiber grown on short rotations. In: *Applied Polymer Symposium No. 28,* J. Wiley and Sons, 393-401.

Stephens, G.L. and Webster, P.J., 1981. Clouds and climate: sensitivity of simple systems. *Journal of the Atmospheric Sciences,* 38, 235-247.

Stommel, H., 1979. Determination of water mass properties of water pumped down from the Ekman layer to the geostrophic flow below. *Proceedings of the National Academy of Sciences,* 76, 3051-3055.

Stuiver, M., 1978. Atmospheric carbon dioxide and carbon reservoir changes. *Science,* 199, 253-258.

Takahashi, T., 1979. Carbon dioxide chemistry in ocean water. In: *Workshop on the Global Effects of Carbon Dioxide from Fossil Fuels,* Ed. W.P. Elliott and L. Machta, US Dept. of Energy, 63-71.

Tans, P.P., 1978. *Carbon-13 and Carbon-14 in Trees and the Atmospheric CO_2 Increase.* Thesis Rijsuniversiteit te Groningen, The Netherlands.

Tellings, M.M.J., 1978. *The Influence of Temperature on Industry Demand for Petroleum Products and Different Types of Energy.* Supply Oil Programmes and Plans, Shell Centre, UK.

Thompson, S.L. and Schneider, S.H., 1979. A seasonal zonal energy balance climate model with an interactive lower layer. *Journal of Geophysical Research,* 84, 2401-2414.

Thorpe, T.E., 1867. On the amount of carbonic acid contained in sea-air. *Quarterly Journal of the Chemical Society of London,* 20, 189-199.

UNESCO, 1980. The carbon budget of the oceans, Report of SCOR Working Group 62, UNESCO, Paris, 12-13 November 1979. *UNESCO Technical Papers in Marine Science,* 34.

US Department of Energy, 1981. *Some Aspects of the Role of the Shallow Ocean in Global Carbon Dioxide Uptake, Report of a Workshop Atlanta, Georgia, 20-22 March 1980,* Ed. R.M. Garrels and F.T. Mackenzie.

Viecelli, J.A., Ellsaesser, H.W. and Burt, J.E., 1981. A carbon cycle model with latitude dependence. *Climatic Change,* 3, 281-302.

Vonder Haar, T.H. and Oort, A.H., 1973. New estimate of annual poleward energy transport by northern hemisphere oceans. *Journal of Physical Oceanography,* 3, 169-172.

Wagener, K., 1978. Total anthropogenic CO_2 production during the period 1800-1935 from carbon-13 measurements in tree rings. *Radiation and Environmental Biophysics,* 15, 101-111.

Walsh, J.J., Rowe, G.T., Iverson, R.L. and McRoy, C.P., 1981. Biological export of shelf carbon is a sink of the global CO_2 cycle. *Nature,* 291, 196-201.

Washington, W.M., Semtner, A.J., Meehl, G.A., Knight, D.J. and Mayer, T.A., 1980. A general circulation experiment with a coupled atmosphere, ocean and sea ice model. *Journal of Physical Oceanography,* 10, 1887-1908.

Watts, R.G., 1980. Discussion of 'Questions concerning the possible influence of anthropogenic CO_2 on atmospheric temperature'. *Journal of Applied Meteorology,* 19, 494-495.

Watts, R.G., 1982. Further discussion of 'Questions concerning the possible influence of anthropogenic CO_2 on atmospheric temperature'. *Journal of Applied Meteorology*, 21, 243-247.

Webb, E.K., Pearman, G.I. and Leuning, R., 1980. Correction of flux measurements for density effects due to heat and water vapour transfer. *Quarterly Journal of the Royal Meteorological Society*, 106, 85-100.

Weiss, R.F., 1974. Carbon dioxide in water and seawater: the solubility of a non-ideal gas. *Marine Chemistry*, 2, 203-215.

Wesely, M.L., Cook, D.R., Hart, R.L. and Williams, R.M., 1982. Air-sea exchange of CO_2 and evidence for enhanced upward fluxes. *Journal of Geophysical Research*, 87, 8827-8832.

Wetherald, R.T. and Manabe, S., 1981. Influence of seasonal variation upon the sensitivity of a model climate. *Journal of Geophysical Research*, 86, 1194-1204.

Wigley, T.M.L., 1982. Energy production and climatic change: an assessment. In: *Uranium and Nuclear Energy: 1981*, Butterworth, 291-322.

Wigley, T.M.L., 1983. The pre-industrial carbon dioxide level. *Climatic Change*, in press.

Wigley, T.M.L., Jones, P.D. and Kelly, P.M., 1980. Scenario for a warm, high-CO_2 world. *Nature*, 283, 17-21.

Williams, J., 1980. Anomalies in temperature and rainfall during warm Arctic seasons as a guide to the formulation of climatic scenarios. *Climatic Change*, 2, 249-266.

WMO/ICSU/UNEP, 1981. *On the Assessment of the Role of CO_2 on Climate Variations and their Impact, Meeting of Experts, Villach, Austria, Novermber 1980*. WCP, Geneva.

Wollast, R., Garrels, R.M. and Mackenzie, F.T., 1980. Calcite-seawater reactions in ocean surface waters. *American Journal of Science*, 280, 831-848.

Wong, C.S., 1978. Atmospheric input of carbon dioxide from burning wood. *Science*, 200, 197-200.

Woodwell, G.M. and Houghton, R.A., 1977. Biotic influences on the world carbon budget. In: *Global Chemical Cycles and their Alterations by Man*, Ed. W. Stumm, Dahlem Konferenzen, 61-72.

Woodwell, G.M., Whittaker, R.H., Reiners, W.A., Likens, G.E., Delwiche, C.C. and Botkin, D.B., 1978. The biota and the world carbon budget. *Science*, 199, 141-146.

Index

Aerosols, effect on climatic warming 99-100
Agriculture, influence of higher CO_2 concentrations on 32
effect of expansion on atmospheric CO_2 35-36
Airborne fraction of excess CO_2 5-7, 21, 23, 41-42, 93-94
Air-sea exchange of CO_2 18-20
Alaska, Point Barrow, record of CO_2 concentration 3-5, 7-8
Albedo, of cloud 67-70
see also Ice-albedo feedback
Alkalinity of seawater 14, 17, 28
Anthropogenic CO_2, global budget 43-45
man-made input to atmosphere 9-11
see also Fossil fuel combustion
Aragonite, supersaturation in surface waters 29
Atmosphere, ocean coupled general circulation models 71-72, 83-84
Atmospheric CO_2 increase, future timescale 93-96
Atmospheric measurements of CO_2 3-9
Atmospheric infrared absorption 50-52

Biological productivity 102-103
Biological uptake of CO_2 in the oceans 27-28
Biosphere, terrestrial, as a source/sink for atmospheric CO_2 7-9, 31-42
indicators of net effect on atmospheric CO_2 36-41
modelling 35-36
role in the anthropogenic CO_2 cycle 41
see also Carbon-13, crops, forests, land-use change, photosynthesis, seasonal variation
Blackbody radiative emission of Earth & atmosphere 50-51
Bomb-produced radiocarbon, as an oceanic tracer 20-21, 24-25
Borate chemistry of seawater 14
Box models of the ocean 21-26
Buffer factor 15-18, 89

Calcite, supersaturation in surface waters 29
Calcium carbonate minerals in the ocean 28-29
Carbon-13, as an indicator of net effect of terrestrial biosphere on atmospheric CO_2 39-41
Carbon-14, balance in the atmosphere-ocean system 23
dilution in atmosphere by fossil fuel CO_2 40
as an oceanic tracer 20-21, 23-24
Carbonate chemistry of seawater 13-18
Carbonate alkalinity 14
Carbonate minerals in seawater 28-29
Cement production, effect on atmospheric CO_2 5, 9-11, 37, 41-42
Charcoal production by biomass burning 36
Chlorofluorocarbons, effects on climatic warming 99-101
Cloud-radiation feedback 68-69, 74
Clouds, effect on the Earth's radiation budget 67
representation in global climate models 67-70, 74-78
response to increased atmospheric CO_2 68
Convection, representation in climate models 61
Convection adjustment 61
Convective mixing in the ocean 20
Coupling of atmospheric & oceanic general circulation models 71-72, 83-84
Crops, effect of elevated CO_2 on growth 31-33, 39

Detection of CO_2-induced climatic change 83, 96-101
Diffusive mixing in the oceans 20, 23-24

Eddy correlation technique for determing air-sea gas exchange 30
Ekman pumping 20
Electricity, effect of climatic change on demand 91-93
Energy, conservation 95
effect of climatic change on demand 92-93
efficiency 101-102
Energy balance models of climate 62-63
Equilibrium capacity of seawater to absorb CO_2 13-18
Equilibrium climate, statistical 66
Evaporation, effects of increased CO_2 upon 78-82

Fertilization effect of elevated CO_2 concentration 31-33, 36
Flue gases, removal of CO_2 from 103-104

Forests, carbon isotope studies 39-41
effect, of clearing & regrowth on atmospheric CO_2 33-35, 39
planting to reduce CO_2 increase 101-103, 105
Fossil fuel combustion 5-11, 21, 28, 34, 36, 37-43, 93-96, 102
Fraction, airborne, of excess CO_2 5-7, 21, 23, 41-42

Gas, effect of climatic change on demand 91-93
Gas exchange rates at air-sea interface 19
General circulation models 65-86
application to the CO_2 problem 73-86
introduction to 65-67
representation of clouds in 67-70
representation of oceans in 70-72
GEOSECS data, air-sea CO_2 exchange 19
oceanic mixing 21, 24-25
Global budget of CO_2 43-45
Global climate models *see* General circulation models
Global production of CO_2 105-106
Greenhouse effect 49-57, 60-63, 68, 99-101
Gross primary production 102

Heat transport by the oceans 71
Henry's Law 13, 17
Historical analogues of warm climates 87-88
Historical CO_2 concentration in the atmosphere 3-9, 88-90
Horticulture, effect of enhanced CO_2 concentrations on 31-33

Ice-cores, analysis for past CO_2 concentrations 88-90
Ice-albedo feedback on temperature 59, 62, 75, 77, 86, 89
Infrared absorption in the atmosphere 50-52
Infrared radiative transfer 52-54
International research on the CO_2/climate problem 105-107
Isopycnal mixing in the oceans 20
Isotopes, of carbon *see* Carbon-13, carbon-14

Land-use change, effect on atmospheric CO_2 33-36
Lag, in climate response due to oceans 71, 83-85

Magnesium calcite minerals, solubility of 29
Man-made CO_2, input to the atmosphere 9-11
global budget of 43-45
see also Fossil fuel combustion
Marine biology 27-28
Mauna Loa, record of atmospheric CO_2 concentration 3-8, 37-39, 45, 109
Measurements of atmospheric CO_2 concentration 3-9
Mesoscale eddies 20
Methane, effect on climatic warming 99-101
Minerals, marine, reaction of CO_2 with 28-29
Mixed-layer model of the ocean 70, 83
Mixing, oceanic 20-26
Model of biosphere 35-36
Models of climate, application to the CO_2 problem 59-63, 66, 68-71, 73-86
coupled atmosphere-ocean 71-72
energy balance, latitudinally varying 62-63
general circulation, three dimensional 65-86
horizontally averaged, vertical column 59-61
one-dimensional 59-63
radiative-convective 59-61, 96-99
sensitivity studies 59-61, 66, 72
summary of CO_2 studies 85-86
transient response to increasing CO_2 83-85
Models of the ocean, for CO_2 uptake 21-26
for climate studies 70-72, 83-85
Monoethanolamine, as a scrubbing fluid for CO_2 removal 104

Natural gas flaring, as a source of CO_2 9
Net ecosystem production 102
Net primary production 102
Nitrogen oxides, effect on climatic warming 99-101
Nuclear power generation, as a non-producer of CO_2 95, 102, 104-105

Observations of atmospheric CO_2 concentration 3-9
Ocean, deep water formation 20, 24
heat transport 71
mixing 20-26
models of CO_2 uptake 21-26
representation in climate models 70-72, 83-85
thermal capacity 70
see also Oceanic uptake of CO_2

Ocean-atmosphere coupled general circulation models 71-72, 83-85
Ocean-atmosphere gas exchange 18-20
Oceanic uptake of CO_2 13-30
air-sea exchange 18-20
biological 27-28
equilibrium capacity 13-18
experimental estimation of 29-30
mixing processes 20-26
models 21-26
reaction with marine minerals 28-29
Oil, effect of climatic change on demand 91-93
Ozone, effect on climatic warming 99-101

Parametrization of subgrid-scale processes in climate models 66, 68
pH of seawater 14, 17, 28
Photochemistry of tropospheric trace gases 99
Photosynthesis 7, 31-33, 39, 102-103
Pioneer settlement 35
Precipitation, effects of increase CO_2 on 78-82
Production of CO_2, by world segments 105-106

Radiative balance of the Earth-atmosphere system 50
Radiative influence of increased CO_2, direct 54-57, 60
Radiative-convective models 59-61, 96-99
Radiocarbon *see* Carbon-14
Rainfall, effect of increased CO_2 on 78-82
Renewable energy sources, as non-producers of CO_2 95, 102, 104-105
Research, international, on the CO_2/climate problem 105-107
Revelle Factor 15-18

Scrubbing of flue gases, for CO_2 removal 103-104
Sea-ice, simulation in climate models 77-78
Sea-surface temperatures, use of climatologically observed in climate models 70, 75
Seasonal variation of atmospheric CO_2 7-8
  as an indicator of net effect of terrestrial biosphere 37-39
Sector model 74
Soil carbon, effect of land-use changes 33-36
Soil moisture, effect of increased CO_2 on 78-82
Solar luminosity, effect on climatic change 96-97
Solar radiation, as a component of the radiation budget 50-52
South Pole, record of CO_2 concentration 3-5, 7-8
Statistical equilibrium climate 66
Suess effect 40
Swamp model of the oceans 70, 74-75
Synthetic fuels, effect on CO_2 increase 93-94

Technological options for reducing rate of CO_2 increase 101-105
Temperature, effect of increased CO_2 on 59-62, 74-79, 83-86
  effect of increases in other trace gases on 99-101
Terrestrial biosphere *see* Biosphere, terrestrial
Thermocline ventilation 20, 24
Timescale of atmospheric CO_2 increase 93-96
Titration alkalinity 14
Total inorganic carbon in seawater 14, 30
Trace gases, effect on climatic warming 99-101
Transient climatic response to increasing CO_2 83-85
Tree-ring carbon-13 studies 40-41
Trees, effect of clearing & regrowth on atmospheric CO_2 33-36, 39
  planting to reduce CO_2 increase 102-103, 105

Vegetation, effect on CO_2 levels 31-36, 39-40
Volcanic dust, effects on climate 96-97

Water budget, effect of increased CO_2 on 78-82
Water vapour, feedback on temperature 52, 59-61, 75
  spectral overlap with CO_2 57
Weighting function 54-56
World production of CO_2 105-106